Cell-Free Synthetic Biology

Cell-Free Synthetic Biology

Special Issue Editor

Seok Hoon Hong

MDPI • Basel • Beijing • Wuhan • Barcelona • Belgrade

MDPI

Special Issue Editor
Seok Hoon Hong
Illinois Institute of Technology
USA

Editorial Office
MDPI
St. Alban-Anlage 66
4052 Basel, Switzerland

This is a reprint of articles from the Special Issue published online in the open access journal *Methods and Protocols* (ISSN 2409-9279) in 2019 (available at: https://www.mdpi.com/journal/mps/special_issues/Cell_free_Synthetic_Biology).

For citation purposes, cite each article independently as indicated on the article page online and as indicated below:

LastName, A.A.; LastName, B.B.; LastName, C.C. Article Title. *Journal Name* **Year**, *Article Number*, Page Range.

ISBN 978-3-03928-022-3 (Pbk)
ISBN 978-3-03928-023-0 (PDF)

Cover image courtesy of Dr. Javin P. Oza, California Polytechnic State University.

Contents

About the Special Issue Editor

Seok Hoon Hong, Dr., has been an Assistant Professor at the Illinois Institute of Technology since 2015. His research interests are developing cell-free protein synthesis platforms and applying synthetic biology tools to control multidrug-resistant bacteria and biofilms. Before joining the Illinois Institute of Technology, he worked on cell-free synthetic biology as a postdoctoral researcher at Northwestern University for four years. He studied the production of novel protein-based materials via site-specific incorporation of non-standard amino acids and advanced genome engineering. His cell-free protein synthesis platform significantly improved the yield of modified proteins and enabled the production of sequence-defined biopolymers. In 2011, he received his Ph.D. at Texas A&M University, studying bacterial biofilms and persister cell formation by protein engineering and synthetic biology. He demonstrated biofilm displacement via a population-driven genetic switch coupled to engineered biofilm dispersal proteins. As of November 2019, his work has led to twenty-eight scientific papers in peer-reviewed journals, one patent, and over sixty presentations at scientific and engineering research conferences.

*methods
and
protocols*

MDPI

Editorial

"Cell-Free Synthetic Biology": Synthetic Biology Meets Cell-Free Protein Synthesis

Seok Hoon Hong

Department of Chemical and Biological Engineering, Illinois Institute of Technology, Chicago, IL 60616, USA;
shong26@iit.edu; Tel.: +1-312-567-8950

Received: 30 September 2019; Accepted: 1 October 2019; Published: 8 October 2019

Since Nirenberg and Matthaei used cell-free protein synthesis (CFPS) to elucidate the genetic code in the early 1960s [1], the technology has been developed over the course of decades and applied to studying both fundamental and applied biology [2]. Cell-free synthetic biology integrating CFPS with synthetic biology has received attention as a powerful and rapid approach to characterize and engineer natural biological systems. The open nature of cell-free (or in vitro) biological platforms compared to in vivo systems brings an unprecedented level of control and freedom in design [3]. This versatile engineering toolkit has been used for debugging biological networks, constructing artificial cells, screening protein libraries, prototyping genetic circuits, developing biosensors, producing metabolites, and synthesizing complex proteins including antibodies, toxic proteins, membrane proteins, and novel proteins containing nonstandard (unnatural) amino acids. The *Methods and Protocols* "Cell-Free Synthetic Biology" Special Issue consists of a series of reviews, protocols, benchmarks, and research articles describing the current development and applications of cell-free synthetic biology in diverse areas.

Although interest in CFPS is growing, new users often face technical and functional issues in choosing and executing the CFPS platform that best suits their needs. An extensive review article by Gregorio et al. [4] provides a guide to help new users overcome the barriers to implementing CFPS platforms in research laboratories. CFPS platforms derived from diverse microorganisms and cell lines can be divided into two categories, including high adoption and low adoption platforms, by clarifying the similarities and differences among cell-free platforms. Various applications have been achieved by using each of these platforms. The authors also review methodological differences between platforms and the instrumental requirements for their preparation. New users can determine which type of cell-free platform could be used for their needs.

Another review article by Jeong et al. [5] summarizes the use of cell-free platforms for engineering synthetic biological circuits and systems. Because synthetic biological systems have become larger and more complex, deciphering the intricate interactions of synthetic systems and biological entities is a challenging task. Cell-free synthetic biology approaches can facilitate rapid prototyping of synthetic circuits and expedite the exploration of synthetic system designs beyond the confines of living organisms. Cell-free platforms can also provide a suitable platform for the development of DNA nanostructures, riboregulators, and artificial cells, and can enable validation of mathematical models for understanding biological regulation.

Incorporating nonstandard amino acids into proteins is an important technology to improve the understanding of biological systems as well as to create novel proteins with new chemical properties, structures, and functions. Improvements in CFPS systems have paved the way to accurate and efficient incorporation of nonstandard amino acids into proteins [6]. Gao et al. [7] describe a rapid and simple method to synthesize unnatural proteins in a CFPS system based on *Escherichia coli* crude extract by using an unnatural orthogonal translational machinery. This protocol provides a detailed procedure for using a CFPS system to synthesize unnatural proteins on demand.

In CFPS systems, the activity of the crude extract is crucial to ensure high-yield protein synthesis and to minimize batch-to-batch variations in the cell-free reaction. Kim et al. [8] describe a practical method for the preparation and optimization of crude extract from genomically engineered *E. coli* strains [9]. This protocol summarizes entire steps of CFPS from cell growth to harvest, from cell lysis to dialysis, and from cell-free reaction setup to protein quantification. Of note, this method can be easily applied to other commercially available or laboratory stock *E. coli* strains to produce highly active crude extracts.

Because CFPS does not use living cells, toxic proteins can be produced in CFPS at high yield. Jin et al. [10] report that colicins, antimicrobial toxins, can be synthesized and optimized through CFPS at high-yield and activity. Chaperone-enriched *E. coli* extracts significantly enhance the protein solubility. Further modification of the system, such as by including the immunity protein that binds to the colicin, improves the cytotoxic activity of colicin. This study demonstrates that CFPS is a viable platform for optimal production of toxic proteins.

Another optimization of CFPS systems by Yang et al. [11] is applied to produce biosimilar therapeutics. Posttranslational modification of mammalian proteins in prokaryotic systems is challenging. However, producing an active form of tissue plasminogen activator containing 17 disulfide bonds can be achieved in an *E. coli*-based CFPS by overexpressing or supplementing with disulfide bond isomerase and optimizing the buffer conditions during the reaction. This study represents an important step toward the development of *E. coli*-based CFPS technology for rapid, inexpensive, on-demand production of biotherapeutics.

Eukaryotic CFPS systems can serve as alternative production systems for mammalian proteins that exhibit insufficient protein folding or posttranslational modification in prokaryotic CFPS systems. Thoring et al. [12] demonstrate that eukaryotic cell-free systems based on eukaryotic lysates have the potential to produce druggable protein targets. WNT proteins and the cytosolically produced hTERT enzyme have been produced and optimized in eukaryotic cell-free systems. The improvement of eukaryotic CFPS platforms has the potential to accelerate drug development pipelines.

In addition to protein production, cell-free systems provide great benefits in advancing metabolic engineering. Lim and Kim [13] review recent developments and prospects of cell-free metabolic engineering, which, in comparison to cell-based metabolic processes, has the benefits of operational simplicity, high conversion yield and productivity, and no environmental release of engineered microorganisms. This article summarizes the importance of configuring cell-free enzyme synthesis and establishing cell-free metabolic engineering in the development of directly programmable metabolic engineering platforms.

I believe that the collection of articles in the "Cell-Free Synthetic Biology" Special Issue of *Methods and Protocols* will provide researchers with both a comprehensive understanding of diverse aspects of cell-free synthetic biology and practical methods to apply cell-free synthetic biology tools and knowledge to advance their studies.

Funding: This work was supported by the National Institute of Allergy and Infectious Diseases of the National Institutes of Health (R15AI130988).

Conflicts of Interest: The author declares no conflicts of interest.

References

1. Nirenberg, M.W.; Matthaei, J.H. The dependence of cell-free protein synthesis in *E. coli* upon naturally occurring or synthetic polyribonucleotides. *Proc. Natl. Acad. Sci. USA* **1961**, *47*, 1588–1602. [CrossRef] [PubMed]
2. Carlson, E.D.; Gan, R.; Hodgman, C.E.; Jewett, M.C. Cell-free protein synthesis: Applications come of age. *Biotechnol. Adv.* **2012**, *30*, 1185–1194. [CrossRef] [PubMed]
3. Perez, J.G.; Stark, J.C.; Jewett, M.C. Cell-free synthetic biology: Engineering beyond the cell. *Cold Spring Harb. Perspect. Biol.* **2016**, *8*, a023853. [CrossRef] [PubMed]
4. Gregorio, N.E.; Levine, M.Z.; Oza, J.P. A user's guide to cell-free protein synthesis. *Methods Protoc.* **2019**, *2*, 24. [CrossRef] [PubMed]

5. Jeong, D.; Klocke, M.; Agarwal, S.; Kim, J.; Choi, S.; Franco, E.; Kim, J. Cell-free synthetic biology platform for engineering synthetic biological circuits and systems. *Methods Protoc.* **2019**, *2*, 39. [CrossRef] [PubMed]

6. Hong, S.H.; Kwon, Y.-C.; Jewett, M.C. Non-standard amino acid incorporation into proteins using *Escherichia coli* cell-free protein synthesis. *Front. Chem.* **2014**, *2*, 34. [CrossRef] [PubMed]

7. Gao, W.; Bu, N.; Lu, Y. Efficient incorporation of unnatural amino acids into proteins with a robust cell-free system. *Methods Protoc.* **2019**, *2*, 16. [CrossRef] [PubMed]

8. Kim, J.; Copeland, C.E.; Padumane, S.R.; Kwon, Y.C. A crude extract preparation and optimization from a genomically engineered *Escherichia coli* for the cell-free protein synthesis system: Practical laboratory guideline. *Methods Protoc.* **2019**, *2*, 68. [CrossRef] [PubMed]

9. Martin, R.W.; Des Soye, B.J.; Kwon, Y.-C.; Kay, J.; Davis, R.G.; Thomas, P.M.; Majewska, N.I.; Chen, C.X.; Marcum, R.D.; Weiss, M.G.; et al. Cell-free protein synthesis from genomically recoded bacteria enables multisite incorporation of noncanonical amino acids. *Nat. Commun.* **2018**, *9*, 1203. [CrossRef] [PubMed]

10. Jin, X.; Kightlinger, W.; Hong, S.H. Optimizing cell-free protein synthesis for increased yield and activity of colicins. *Methods Protoc.* **2019**, *2*, 28. [CrossRef]

11. Yang, S.-O.; Nielsen, G.H.; Wilding, K.M.; Cooper, M.A.; Wood, D.W.; Bundy, B.C. Towards on-demand *E. coli*-based cell-free protein synthesis of tissue plasminogen activator. *Methods Protoc.* **2019**, *2*, 52. [CrossRef]

12. Thoring, L.; Zemella, A.; Wüstenhagen, D.; Kubick, S. Accelerating the production of druggable targets: Eukaryotic cell-free systems come into focus. *Methods Protoc.* **2019**, *2*, 30. [CrossRef] [PubMed]

13. Lim, H.J.; Kim, D.-M. Cell-free metabolic engineering: Recent developments and future prospects. *Methods Protoc.* **2019**, *2*, 33. [CrossRef] [PubMed]

methods and protocols

MDPI

Review

A User's Guide to Cell-Free Protein Synthesis

Nicole E. Gregorio [1,2], Max Z. Levine [1,3] and Javin P. Oza [1,2,*]

1   Center for Applications in Biotechnology, California Polytechnic State University, San Luis Obispo, CA 93407, USA; negregor@calpoly.edu (N.E.G.); mzlevine@calpoly.edu (M.Z.L.)
2   Department of Chemistry and Biochemistry, California Polytechnic State University, San Luis Obispo, CA 93407, USA
3   Department of Biological Sciences, California Polytechnic State University, San Luis Obispo, CA 93407, USA
*   Correspondence: joza@calpoly.edu; Tel.: +1-805-756-2265

Received: 15 February 2019; Accepted: 6 March 2019; Published: 12 March 2019

Abstract: Cell-free protein synthesis (CFPS) is a platform technology that provides new opportunities for protein expression, metabolic engineering, therapeutic development, education, and more. The advantages of CFPS over in vivo protein expression include its open system, the elimination of reliance on living cells, and the ability to focus all system energy on production of the protein of interest. Over the last 60 years, the CFPS platform has grown and diversified greatly, and it continues to evolve today. Both new applications and new types of extracts based on a variety of organisms are current areas of development. However, new users interested in CFPS may find it challenging to implement a cell-free platform in their laboratory due to the technical and functional considerations involved in choosing and executing a platform that best suits their needs. Here we hope to reduce this barrier to implementing CFPS by clarifying the similarities and differences amongst cell-free platforms, highlighting the various applications that have been accomplished in each of them, and detailing the main methodological and instrumental requirement for their preparation. Additionally, this review will help to contextualize the landscape of work that has been done using CFPS and showcase the diversity of applications that it enables.

Keywords: cell-free protein synthesis (CFPS); in vitro transcription-translation (TX-TL); cell-free protein expression (CFPE); in vitro protein synthesis; cell-free synthetic biology; cell-free metabolic engineering (CFME)

---

1. Introduction

Cell-free protein synthesis (CFPS) emerged about 60 years ago as a platform used by Nirenberg and Matthaei to decipher the genetic code and discover the link between mRNA and protein synthesis [1]. Since this discovery, the CFPS platform has grown to enable a variety of applications, from functional genomics to large-scale antibody production [2,3]. Currently, CFPS has been implemented using cell extracts from numerous different organisms, with their unique biochemistries enabling a broad set of applications. In an effort to assist the user in selecting the CFPS platform that is best suited to their experimental goals, this review provides an in-depth analysis of high adoption CFPS platforms in the scientific community, the applications that they enable, and methods to implement them. We also review applications enabled by low adoption platforms, including applications proposed in emerging platforms. We hope that this will simplify new users' choice between platforms, thereby reducing the barrier to implementation and improving broader accessibility of the CFPS platform.

The growing interest in CFPS is the result of the key advantages associated with the open nature of the platform. The CFPS reaction lacks a cellular membrane and a functional genome, and consequently is not constrained by the cell's life objectives [4]. Therefore, the metabolic and cytotoxic burdens placed on the cell when attempting to produce large quantities of recombinant proteins in vivo are obviated in CFPS [5]. The CFPS platform is amenable to direct manipulation of the environment of protein production because it is an open system (Figure 1). In some cases higher protein titers can be achieved using CFPS because all energy in the system is channeled toward producing the protein of interest (Figure 2) [6]. Moreover, CFPS reactions are flexible in their setup, allowing users to utilize a variety of reaction formats, such as batch, continuous flow, and continuous exchange, in order to achieve the desired protein titer (Figure 3). These advantages make CFPS optimally suited for applications such as the production of difficult-to-synthesize proteins, large proteins, proteins encoded by high GC content genes, membrane proteins, and virus-like particles (Figures 4A and 5A). The scalable nature of CFPS allows it to support the discovery phase through high-throughput screening as well as the production phase through large-scale biomanufacturing. Additional high impact applications include education, metabolic engineering, and genetic code expansion.

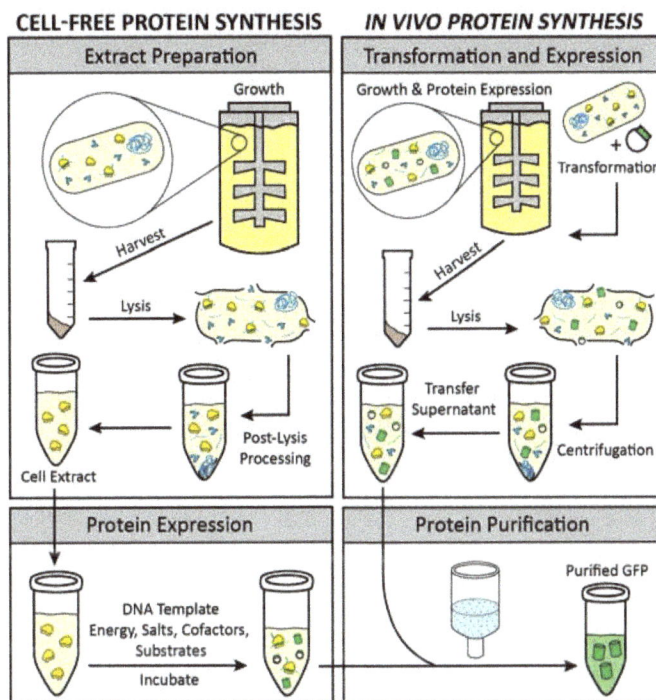

Figure 1. A comparison of cell-free and in vivo protein synthesis methods. Through visualization of the main steps of in vitro and in vivo protein expression, the advantages of cell-free protein synthesis emerge. These include the elimination of the transformation step, an open reaction for direct manipulation of the environment of protein production, the lack of constraints based on the cell's life objectives, the channeling of all energy toward production of the protein of interest, and the ability to store extracts for on-demand protein expression. Green cylinders represent synthesized green fluorescent protein (GFP).

While the number of cell-free platforms based on different organisms has grown substantially since its conception, the basic steps for successful implementation of a cell-free platform are analogous across platforms (Figure 6). In brief, users must culture the cell line of interest from which transcription and translation machinery are to be extracted. Next, the user must lyse the cells while maintaining ribosomal activity in the lysate, prepare cell extract by clarifying the lysate through various methods, and then utilize the prepared cell extract in CFPS reactions to synthesize the protein of interest.

These basic steps have many nuanced variations from platform to platform, and even within platforms. Lysis methods in particular are extremely variable and commonly used methods include homogenization, sonication, French press, freeze thaw, nitrogen cavitation, bead beating [7]. Extract preparation varies by centrifugation speeds, run off reactions, dialysis, or treatment with nucleases to remove endogenous DNA or RNA. Here, we report methodologies used most commonly for obtaining highest volumetric yields of the target protein (Tables 1–3). We also report low adoption platforms including emerging platforms that adapt these methods for continued innovations in CFPS.

Based on nearly 60 years of literature, we have divided CFPS platforms into two categories: high adoption and low adoption platforms. The latter also includes emerging platforms. High adoption platforms for CFPS are based on extracts from the following cell lines: *Escherichia coli*, *Spodoptera frugiperda* (insect), *Saccharomyces cerevisiae* (yeast), Chinese hamster ovary, rabbit reticulocyte lysate, wheat germ, and HeLa cells. These platforms have been well optimized and utilized since their conception and are most easily implemented by new users due to the breadth of supporting literature (Figure 4). Platforms that have experienced low adoption to date include *Neurospora crassa*, *Streptomyces*, *Vibrio natriegens*, *Bacillus subtilis*, Tobacco, *Arabidopsis*, *Pseudomonas Putida*, *Bacillus megaterium*, Archaea, and *Leishmania tarentolae*. These platforms have not been widely used or developed, and some have even emerged in the last two years as promising candidates for new applications (Figure 5). Trends in CFPS literature demonstrate that there is continued development and optimization of platforms, and the emerging platforms are likely to be the source of rapid innovations. We also anticipate significant development toward the broad dissemination and utilization of CFPS platforms.

2. CFPS Reaction Formats

As an open and highly personalized platform, CFPS reactions can be executed in a variety of formats, including coupled, uncoupled, batch, continuous flow, continuous exchange, lyophilized, or microfluidic formats depending on the needs of the user. Additionally, there are a variety of commercial CFPS kits available for users looking to implement CFPS quickly, without the need for long-term or large-scale usage. Here we describe the differences and utility of each format.

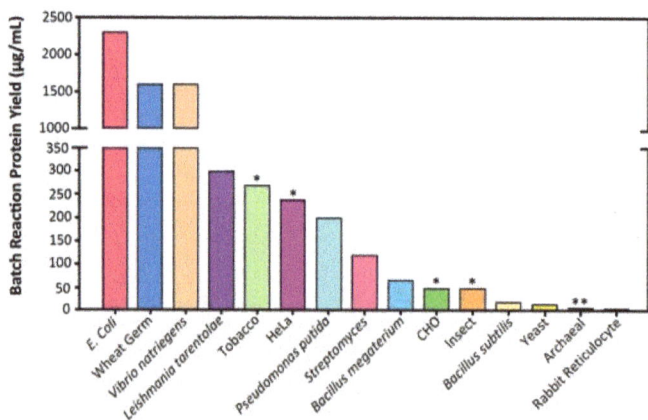

Figure 2. Comparison of protein yields across cell-free platforms. The volumetric yield of each platform is reported for batch reactions producing GFP. Platforms that report volumetric yield for reporter proteins luciferase (*) or ChiAΔ4 (**) are indicated. Information for batch mode protein yields of the *Arabidopsis* and *Neurospora crassa* platforms was not found. Yields were obtained from the following sources: *E. coli* [8], wheat germ [9], *Vibrio natriegens* [10], *Leishmania tarentolae* [11], tobacco [12], HeLa [13], *Pseudomonas putida* [14], *Streptomyces* [15], *Bacillus megaterium* [16], Chinese hamster ovary [17], insect [18], *Bacillus subtilis* [16], yeast [19], archaeal [20], and rabbit reticulocyte [21].

2.1. Commercial Systems

Many of the high adoption CFPS platforms have been commercialized as kits available for users to quickly leverage the advantages of CFPS for their research. This has generally been the best option for labs lacking the access and technical expertise necessary to produce their own cell extracts. Commercial kits enable users to implement CFPS easily, but for extensive usage, they may not be cost-effective. For example, in house prepared *E. coli* CFPS costs about $0.019/μL of reaction while commercial lysate-based kits cost $0.15–0.57/μL of reaction [22]. Currently commercial kits exist for *E. coli* (New England Biolabs, Promega, Bioneer, Qiagen, Arbor Biosciences, ThermoFisher, Creative Biolabs), rabbit reticulocyte (Promega, Creative Biolabs), wheat germ (Promega, Creative Biolabs), *Leishmania tarentolae* (Jena Bioscience), insect (Qiagen, Creative Biolabs), Chinese hamster ovary (Creative Biolabs), HeLa (ThermoFisher, Creative Biolabs), and plant cells (LenioBio).

In addition to cell-extract-based CFPS kits, the PURExpress kit is comprised of a reconstitution of purified components of the transcription and translation machinery from *E. coli*. Specifically, the PURE (**p**rotein synthesis **u**sing **r**ecombinant **e**lements) system utilizes individually purified components in place of cell extract. These include 10 translation factors: T7 RNA polymerase, 20 aminoacyl-tRNA synthetases, ribosomes, pyrophosphatase, creatine kinase, myokinase, and nucleoside diphosphate kinase [23,24]. This system requires overexpression and purification of each component but benefits from the absence of proteases and nucleases, and the defined nature of the system. Overall, the PURE system allows for high purity and somewhat easier manipulation of the reaction conditions than even cell-extract-based CFPS [23]. Moreover, if all synthesized components are affinity-tagged, they can be easily removed post-translationally to leave behind the protein of interest [24]. This system may provide advantages for the synthesis of properly folded proteins with supplemented chaperones, genetic code expansion, and display technologies [23–25]. The PURE system would be significantly more time-consuming to produce in-house but is available commercially (New England BioLabs, Creative Biolabs, Wako Pure Chemical Industries). However, these kits are expensive ($0.99/μL of reaction) when compared to both in-house and commercially available extract-based CFPS [22]. They are also significantly less productive (~100 μg/mL) than their extract-based *E. coli* CFPS counterpart (Figure 2) [23,26].

2.2. Coupled and Uncoupled Formats

CFPS reactions can be performed in coupled or uncoupled formats, and the choice is dependent on the platform being used and the user's needs. Coupled reactions allow transcription and translation to take place within a single tube, such that the supplied DNA template can be transcribed into mRNA, which is then translated into protein within a one-pot reaction. The advantage of coupled CFPS is the ease of reaction setup, but it may result in suboptimal yields for eukaryotic platforms. Uncoupled reactions typically consist of an in vitro transcription reaction followed by mRNA purification; the purified transcripts are then supplied to the cell-free translation reaction containing the cell extract for production of the protein of interest. Uncoupled reactions are more often utilized in eukaryotic CFPS platforms due to mRNA processing for more efficient translation of certain transcripts. As an example, pseudouridine modification for mRNA in the rabbit reticulocyte platform has been demonstrated to enhance translation [27]. Uncoupled reactions also allow for different conditions between transcription and translation reactions, which can improve yields [9]. Uncoupled reactions can be achieved in any platform by supplying the reaction with mRNA instead of DNA, but mRNA can be more difficult to handle and does degrade more quickly in the CPFS reaction [28].

2.3. Batch, Continuous Flow, and Continuous Exchange Formats

CFPS reactions can be performed in batch format for simplified setup, or in continuous formats for improved protein yields. Reactions are most easily, quickly, and cheaply set up in batch format because all necessary reactants are added to a single tube and incubated for protein synthesis to occur

(Figure 3). However, the duration of a batch reaction is dependent on the amount of substrate available and the amount of inhibitory byproduct produced, which can result in low yields for some platforms (Figure 2). On the other hand, continuous flow and continuous exchange CFPS reactions utilize a two-chamber system to supply reactants and remove products, for increased reaction duration and higher protein yields [29–32]. In continuous exchange cell-free (CECF), the CFPS reaction is separated from a reactant-rich feed solution via a semi-permeable membrane, such that new reactants move into the reaction and byproducts move out, while the protein product remains in the reaction compartment (Figure 3) [31]. For continuous flow cell-free (CFCF), the feed solution is continuously pumped into the reaction chamber, while the protein of interest and other byproducts are pushed out through an ultrafiltration membrane (Figure 3) [33].

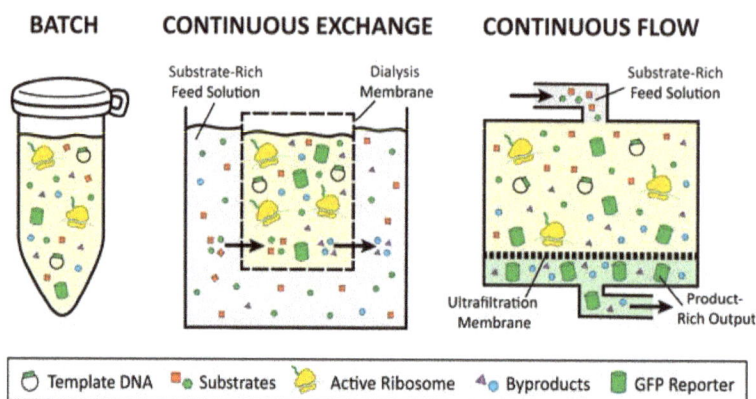

Figure 3. Comparison of batch, continuous flow, and continuous exchange reaction formats. Batch reactions contain all the necessary reactants within a single reaction vessel. Continuous exchange formats utilize a dialysis membrane that allows reactants to move into the reaction and byproducts to move out, while the protein of interest remains in the reaction compartment. Continuous flow formats allow a feed solution to be continuously pumped into the reaction chamber while the protein of interest and other reaction byproducts are filtered out of the reaction.

Batch reactions are well suited to platforms that exhibit high protein yields and to applications that require simple and fast setup (Figure 2). These applications may include high-throughput screening and education. Moreover, batch reactions can be easily scaled up in platforms such as *E. coli* and wheat germ, due to the ability to scale growth and reaction setup linearly. Platforms such as Chinese hamster ovary, yeast, and rabbit reticulocyte, which suffer from low protein yields, may require a CFCF or CECF setup to generate sufficient amounts of protein. Continuous formats have already been successfully constructed in Chinese hamster ovary, insect, *E. coli*, wheat germ, and yeast [30,32,34–37]. For example, continuous formats have allowed for the synthesis of 285 µg/mL of human EGFR to be produced by the insect platform, 980 µg/mL of membrane protein in the Chinese hamster ovary platform, and up to 20,000 µg/mL of protein in wheat germ [9,38,39]. Continuous formats may also be used for large-scale protein synthesis reactions in industrial applications [38,39]. Scale-up of CFPS reactions will be discussed in more detail in Section 3.2.5 titled "Large-Scale."

2.4. Lyophilization

Lyophilization, or freeze-drying, has been used as a technique to stabilize cell extracts for long-term and higher temperature storage, and to provide a condensed format to reduce necessary storage space. By overcoming the cold chain, lyophilization could help enable applications such as on-demand biosensors for diagnostics, therapeutic production in remote locations, personalized medicines, and more [40]. Lyophilization has only been heavily pursued for *E. coli* extract thus far, with some additional work done on the lyophilization of other CFPS reagents and the addition of lyoprotectant additives, and with preliminary work done in wheat germ [41].

Traditionally, aqueous cell-extract is stored at −80 °C, and its activity is reduced by 50% after just one week of storage at room temperature, with all activity lost after a month [42]. In comparison, lyophilized extract maintains approximately 20% activity through 90 days of storage at room temperature. Importantly, the process of lyophilization does not negatively impact reaction yields. A CFPS reaction run directly after lyophilization could achieve the same yields as an aqueous reaction [42]. Lyophilized extract also reduces storage volume to half and mass to about one-tenth [42]. Importantly, the process of lyophilization itself does not negatively impact extract productivity [43]. Lyophilization of extract has also been done on paper, rather than in a tube, to further improve storage and distribution of cell-free technology [44,45].

Some work has been done to test the viability of lyophilizing CFPS reagents necessary for a phosphoenolpyruvate-based reaction setup. These reagents were lyophilized with or without the extract, and while viability was improved over aqueous storage of the reagents at higher temperatures, the combined extract and reagent mixture posed new challenges to the handling of the lyophilized powder due to the resulting texture [42]. Other users have lyophilized the template of interest separately from otherwise fully prepared CFPS reaction for classroom applications, such that the template is simply rehydrated and added to the reaction pellet to begin protein synthesis [46,47]. Additionally, lyoprotectants for cell-free applications have been briefly screened, including sucrose, which provided no obvious benefits to storage stability [42].

2.5. Microfluidics Format

The growing field of microfluidics consists of many broad methodologies that generally involve the manipulation of fluids on the micron scale on devices with critical dimensions smaller than one millimeter [48]. These devices, when paired with cell-free extracts, provide cost-effective and rapid technologies capable of high-throughput assays to generate protein in an automated series of channels that often consist of mixers, reactors, detectors, valves, and pumps on a miniaturized scale [49]. The utilization of microfluidics to pioneer biomedical and diagnostic approaches for sensing and monitoring environmental and health issues has been achieved within *E. coli*, wheat germ, and insect platforms [49]. Examples of applications that utilize the microfluidics format include both the *E. coli* and wheat germ platforms to test for the presence of ricin in orange juice and diet soda through the generation of a reporter protein [50,51]. The insect platform was also used in a Transcription-RNA Immobilization and Transfer-Translation (TRITT) system for the production of a cytotoxic protein with simultaneous non-standard amino acid incorporation for fluorescence labeling [52].

3. Applications of Cell-Free Protein Synthesis

3.1. Introduction to Platform Categorization

In the 60 years since cell-free protein synthesis emerged, a multitude of platforms have been developed based on cell extracts from a variety of organisms. These include extracts from bacterial, archaeal, plant, mammalian, and human cell lines. Each resulting platform varies in ease of preparation, protein yields, and in possible applications resulting from the unique biochemistry of the given organism. In this review, we have divided these various platforms into two categories: high adoption and low adoption (Figures 4 and 5 and Supplementry Materials). The platforms have been categorized based on our understanding of their development and the degree to which they have been adopted by the field, as quantified by the number of peer-reviewed publications that utilize each platform (Figures 4B and 5B). This categorization allows new users to identify platforms that have been best established and to explore the applications that they enable. We believe that the depth of literature available for these platforms makes them optimally suited for newer users. Low adoption platforms may be particularly useful for niche applications, but have not been optimized thoroughly, or are currently emerging in the field. Therefore, these platforms may be more difficult to implement due to minimal development. Platforms with fewer than 25 peer-reviewed publications to date have been categorized as "low adoption."

3.2. High Adoption Platforms

High adoption platforms include those based on *E. coli*, insect, yeast, Chinese hamster ovary, rabbit reticulocyte lysate, wheat germ, and HeLa cells (Figure 4). These platforms have been utilized for a variety of applications and have withstood the test of time to establish their utility and versatility within the CFPS field. Briefly, bacterial CFPS platforms including *E. coli* tend to have higher protein yields and are typically easier and faster to prepare (Figure 2). However, they can be limited in some applications such as post-translational modifications, membrane protein synthesis, and other difficult-to-synthesize proteins. In such cases, eukaryotic platforms are well suited for the synthesis of traditionally difficult proteins without requiring significant augmentation or modifications to the cell extracts. Within the eukaryotic platforms, wheat germ provides the highest productivity; rabbit reticulocyte, Chinese hamster ovary, HeLa, yeast, and insect platforms give significantly lower yields but may have other advantages for post-translation modifications, membrane proteins, or virus-like particles. In order to enable users to select a platform that will support their experimental goals, the discussion of high adoption platforms is application-driven. For each application, the relevant platform and reaction formats are discussed.

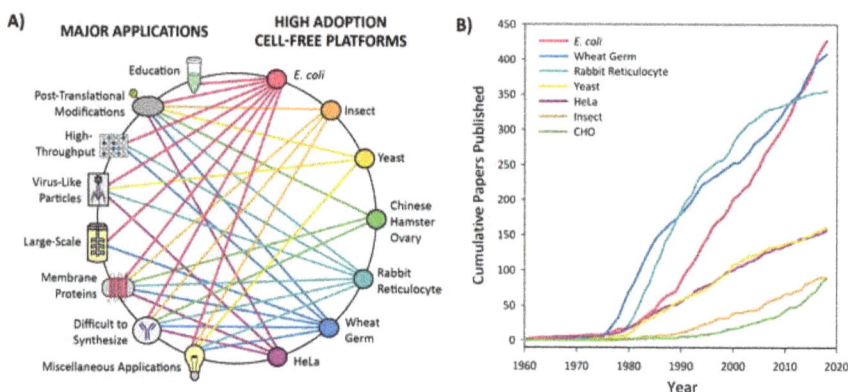

Figure 4. High adoption cell-free platforms and their applications. (**A**) Web of the applications enabled by high adoption cell-free platforms. The connections shown are based on applications that have been published for each respective platform. Applications under "difficult to synthesize proteins" include the production of antibodies, large proteins, ice structuring proteins, and metalloproteins. Miscellaneous applications include studies of translational machinery, genetic circuits, metabolic engineering, and genetic code expansion. (**B**) Cumulative number of peer-reviewed publications over the last 60 years for high adoption platforms. The metric of cumulative publications by platform is used to indicate which platforms are most utilized, with platforms having over 25 papers categorized as high adoption. These data were generated by totaling papers from a PubMed Boolean search of the following: ("cell free protein synthesis" OR "in vitro transcription translation" OR "in vitro protein synthesis" OR "cell free protein expression" OR "tx tl" OR "cell-free translation") AND "platform name." The platform name used for each search corresponds to the name listed in the graph's key. This information was collected on 23 December 2018, and the search results for each platform can be found in Supplemental Information. While this metric is an indicator of the level of adoption for each platform, it does suffer from false positive search results, such as papers reporting studies in which the researchers produce recombinant proteins from the organism of interest rather than from cell extract derived from that organism.

3.2.1. Education

The open nature of the CFPS system and the resulting access to directly manipulate cellular machinery enables inquiry-based learning opportunities that make CFPS particularly suitable for the classroom. The first application of CFPS technology for education is the BioBits kits, which were tested with students of various ages [46,53]. These kits offer versatile experimental options and are relatively inexpensive (about $100 for a class set). The BioBits Bright and Explorer kits represent the diversity of classroom experiments and applications that can be enabled by CFPS, from production of fluorescent proteins, to hydrogel production, and to identification of fruit DNA [46,53]. These possibilities show that CFPS enables inquiry-based learning of concepts in biochemistry in a hands-on fashion. The stability of CFPS classroom kits is achieved through lyophilization of reaction components. More information on lyophilization of CFPS can be found in Section 2.4, titled "Lyophilization."

3.2.2. Post-Translational Modifications (PTMs)

Post-translational modifications (PTMs) can greatly affect protein folding, activity, and stability, which may be essential for therapeutic proteins, membrane proteins, and virus-like particles, among others [54]. As such, the ability to incorporate various post-translational modifications (PTMs) into the protein of interest is a key consideration when choosing a CFPS platform. PTMs achieved through genetic code expansion will be discussed in Section 3.2.8.4 "Genetic Code Expansion." Here, we cover some key PTMs possible in each high adoption platform and the necessary modifications of the platform that may be needed to achieve them. A key consideration is that platforms with endogenous microsomes demonstrate a greater capacity to support PTMs. This makes platforms such as Chinese hamster ovary, HeLa, and insect well-suited for this application, as endogenous microsomes are formed from endoplasmic reticulum and maintained during extract preparation. However, when endogenous microsomes are utilized, a new "black box" is introduced to the system, which limits user control and restricts PTM choice to those innately possible in the cell line [54].

In the rabbit reticulocyte platform, a variety of PTMs have been investigated, including glycosylation, cleavage of signal proteins, prenylation, and disulfide bond formation [55–61]. However, rabbit reticulocyte extract requires the addition of exogenous microsomes for PTM incorporation. The platform has also been used to probe the specificity of signal sequence differences between glycosylphosphatidylinositol anchoring and translocation to the ER lumen, which was found to be sensitive to even single residue changes [56].

Insect cell CFPS, which contains endogenous microsomes, allows for signal peptide cleavage, glycosylation, phosphorylation, N-myristoylation, N-acetylation, prenylation, and ubiquitination [62–70]. These possible PTMs are similar to those of rabbit reticulocyte and other mammalian platforms [62]. Disulfide bond formation can also be achieved in these platforms by preparing the cell extract under non-reducing conditions, and adding glutathione along with protein disulfide isomerase to the reactions [71]. The insect cell-free platform was even used to discover new proteins containing a PTM of interest. These techniques utilized MALDI-TOF MS screening of a library of metabolically labeled cDNA clones with motifs matching N-myristoylated proteins to determine which were most susceptible to this PTM [72].

Some PTMs can be achieved in *E. coli*-based CFPS, but this application is generally more technically challenging due to a lack of endogenous microsomes and the limited number of PTMs possible in bacteria when compared to eukaryotes [64]. However, utilization of *E. coli* remains advantageous in terms of overall protein yields and ease of extract preparation, which have prompted the development of PTMs in this platform. The open nature of the reaction enables users to tune redox conditions to make disulfide bond formation feasible in this platform. Additionally some N-linked glycosylation has been made possible through the supplementation of glycosylation machinery [73]. Glycosylation was first achieved through the addition of purified glycosylation components after completed cell-free translation, which was effective, but relatively time-consuming [74]. More recently, oligosaccharyltransferases have been synthesized in CFPS and shown to be active in in vitro glycosylation without the need for purification [75]. Furthermore, *E. coli* strains that have been optimized for glycoprotein synthesis have been used to prepare cell-free extract, such that glycosylation can be pursued in a one-pot system [54].

Chinese hamster ovary extract contains endogenous microsomes, which provide glycosyltransferases for glycoprotein synthesis, chaperones, and other molecules necessary for disulfide bridge formation [34]. Yeast has also been a platform of interest for protein glycosylation, with glycosylation achieved when a completely homologous system was used and yeast microsomes were added. However, yields in this platform are much lower than in *E. coli* [76]. The wheat germ platform also requires exogenous microsome addition, which has allowed for some PTMs to be incorporated [64,77]. A human-based hybridoma-cell extract platform, similar to that of HeLa cell-based extracts, was able to glycosylate human immunodeficiency virus type-1 envelope protein 120 [78].

3.2.3. High-Throughput Screening

The ability to achieve high-throughput protein production is a major advantage of CFPS, as it enables rapid production and screening of a variety of protein products much faster than in in vivo protein expression (Figure 1). Coupled CFPS allows for DNA templates to be plugged in directly without the need for cell transformation/transfection, and in some cases, assays of protein products can be done without the need for purification, creating a powerful one-pot system [28]. A key application of high-throughput CFPS is functional genomics, which allows for the elucidation of new genes and their corresponding protein function. High-throughput screening can be pursued in any platform, but most often utilizes *E. coli*, wheat germ, and rabbit reticulocyte extracts. Here we will discuss some specific examples of CFPS for high-throughput applications.

The *E. coli* platform has been widely used and is well-developed, with relatively simple extract preparation and high yields making it a prime candidate for high-throughput synthesis (Table 2, Figure 2). One notable application of *E. coli*-based CFPS is the ability to screen antibody mutant libraries in rapid design–build–test cycles for antibody engineering. The best mutants could later be scaled up in the same platform for industrial level synthesis (see Section 3.2.5, titled "Large-Scale") [2]. Additionally, the *E. coli* platform has been used for high-throughput functional genomics to identify numerous gene products involved in complex metabolic systems that result in protein accumulation and folding in vitro [3].

While high-throughput applications commonly utilize *E. coli*, the eukaryotic wheat germ platform has advantages for synthesis of soluble, active protein, making it better suited for structural and functional analysis of certain proteins in CFPS [79]. The wheat germ platform has shown the capacity to perform as a "human protein factory" when it was utilized in an attempt to produce 13,364 human proteins. Using the versatile Gateway vector system to generate entry clones allowed for successful synthesis of 12,996 of the human proteins, with many displaying successful function [80].

CFPS from rabbit reticulocyte extract can also be used in a high-throughput fashion for protein microarrays, in order to study protein function, interaction, and binding specificity [81,82]. Ribosome and mRNA display technologies as well as in vitro compartmentalization are also possible in the rabbit reticulocyte platform and allow for genes to be linked to their protein products for functional genomic studies [83,84]. Lastly, the Chinese hamster ovary platform is a candidate for high-throughput synthesis, but examples of implementation have not been demonstrated to date [34,85].

3.2.4. Virus-Like Particles

Virus-like particles (VLPs) are capsids of viruses lacking genomic material, meaning that they are a highly organized and symmetrical aggregations of proteins, capable of carrying molecules of interest within them. As such, production of VLPs allows for study of viral assembly, the creation of effective vaccines, drug delivery using encapsulation, and materials science applications [86]. While VLPs can be produced in vivo, production in CFPS platforms offers advantages including the ability to synthesize toxic VLPs and to manipulate the redox conditions of the reaction for proper disulfide bond formation, which may be essential for thermal stability [86]. The versatility of the CFPS reaction also allows for a single, more robust platform capable of producing many types of VLPs at scalable, higher yields and with easier modification of reaction setup than would be possible in vivo [87].

A variety of CFPS platforms have been used to produce many different VLPS, including *E. coli*, HeLa, rabbit reticulocyte, and yeast. The *E. coli* platform has been used to optimize disulfide bond formation in Qβ VLPs by expression without change to the redox state of the reaction and subsequent exposure to diamide to form disulfide bonds post-assembly, as VLP formation would occur naturally. The Qβ VLP has also been co-expressed with A2 protein, which naturally occurs in the full virus for infection and competitive inhibition purposes [88]. Additionally, human hepatitis B core antigen was produced by supplementation with disulfide forming agents glutathione and disulfide isomerase [86]. MS2 bacteriophage coat proteins have also been expressed in high yields using *E. coli*-based CFPS [87]. Both MS2 and Qβ VLPs have been produced with non-standard amino acid enabled click chemistry, allowing proteins, nucleic acids, and polymer chains to be attached to the surface of the VLPs [89]. In the last year, the *E. coli* platform has enabled the production of the largest biological entities thus far in a CFPS platform: fully functional T7 and T4 bacteriophages [90].

The HeLa cell-based CFPS platform has been used for poliovirus synthesis [91]. Rabbit reticulocyte CFPS has enabled viral assembly studies of HIV Gag protein assembly, which forms immature but fully spherical capsids in CFPS [92]. Furthermore, adenovirus type 2 fibers are able to self-assemble into trimers in rabbit reticulocyte CFPS reactions and hepatitis C core proteins are able to form into capsids, which is not seen in mammalian cell cultures [93,94]. The yeast platform has allowed for optimization of translation of VLPs such as human papillomavirus 58 (HPV 58). Synthesis of this VLP through CFPS could enable the study of capsid assembly and encapsulation mechanisms for HPV [95,96].

3.2.5. Large-Scale

The demonstrations of implementing CFPS from a high-throughput scale for discovery to a manufacturing scale have expanded the utility of this platform [2,97]. Users interested in leveraging this capacity for applications such as the production of antibodies and industrial enzymes, as well as CFPS kit production for field or educational uses, should consider the technical details of scaling up the entire workflow for CFPS (Figure 6). This begins with the capacity to scale cell growth, as well as scaling extract preparation. Platforms that enable this scalability include *E. coli*, wheat germ, and rabbit reticulocyte (Tables 1 and 2) [97–99]. The insect, yeast, and Chinese hamster ovary platforms may also be amenable to scale-up in culture growth, as they are fermentable, but large-scale extract preparation has not been well studied to date [100].

Next, platforms must have scalable CFPS reactions that maintain volumetric protein yields even in large-scale reactions. *E. coli* CFPS has been shown to scale over many orders of magnitude in batch format, from reactions as small as 10 µL to as large as 100 L [97]. Within this range of reaction sizes, volumetric protein yields remain constant if the proper reaction vessel is used. For example, reactions up to 100 µL can be run in 1.5–2 mL microcentrifuge tubes, while reaction over 100 µL should be run in 24-well microtiter plates or a similar thin-layer format [31,101,102]. For liter-scale reactions, bioreactors and fermenters have been used [2,97]. The importance of vessel size for scale-up of batch reactions is due in part to the need for proper oxygen exchange, such that increasing the surface area to volume ratio of the reaction can significantly improve reaction yields [31,102]. The scalability of the *E. coli* platform and discovery of cost-effective metabolisms makes it well suited for industrial applications, as has been demonstrated by companies such as Sutro Biopharma, who use CFPS to produce large batches of antibodies in vitro [103,104].

The wheat germ platform has been used for reaction scale-up through a robotic discontinuous batch reaction that can perform reactions up to 10 mL in volume. This setup is capable of producing at least 2 mg/mL of the protein of interest, including DCN1, involved in ubiquitination, human sigma-1 receptor, and bacteriorhodopsin transmembrane proteins. This system utilizes multiple cycles of concentration, feed buffer addition, mRNA template addition, and incubation to achieve high protein yields with minimized extract usage, an idea similar to continuous flow cell-free (CFCF) and continuous exchange cell-free (CECF) [30,99]. CECF and CFCF formats may also be used to scale up reaction size and increase protein yields as discussed in Section 2.3 "Batch, Continuous Flow, and Continuous Exchange Formats." Continuous formats have been pursued in Chinese hamster ovary, insect, *E. coli*, wheat germ, and yeast [30,32,34–37]. Overall, the *E. coli* and wheat germ platforms are most amenable to large-scale synthesis, as scale-up of the entire CFPS workflow has been demonstrated.

3.2.6. Membrane Proteins

The study of membrane proteins is an integral component of proteomics due to their high abundance within organisms. Approximately 25% of all sequenced genes code for hydrophobic proteins that integrate themselves into cell membranes [105]. Membrane proteins serve a plethora of functions within cells including cell recognition, immune response, signal transduction, and molecule transport. However, expressing these complete proteins in vivo in their correct conformation often poses a challenge due to the naturally low abundance during expression, high hydrophobicity, the necessity of translocation into the membrane, and the impact to the host cell's membrane integrity.

CFPS platforms are able to circumvent these challenges by avoiding dependence on the structural integrity of the cell membrane via the non-membrane bound system [106]. In addition, the supplementation of microsomes, vesicle-like structures, or the presence of endoplasmic reticulum fragments during extract preparation (endogenous microsomes) allows membrane proteins to correctly fold and incorporate themselves into these structures during protein synthesis. Namely, the HeLa, Chinese hamster ovary, and insect platforms all contain endogenous microsomes formed via rupturing of the endoplasmic reticulum during extract preparation. These platforms have successfully expressed a number of membrane proteins ranging from a two-transmembrane malarial protein (HeLa), to epidermal growth factor receptor proteins (Chinese hamster ovary), and finally to a KcsA potassium channel (insect) [39,107,108].

Platforms that require exogenous addition of microsomal structures for membrane protein expression include rabbit reticulocyte, wheat germ, and *E. coli*. The rabbit reticulocyte platform, with the supplementation of semipermeable cells, has been demonstrated to properly express MHC class I heavy chain membrane proteins in their correct conformations [109]. The wheat germ platform has successfully expressed human, mouse, and mycobacterium desaturase complexes with the addition of liposomes, as well as plant solute transporters, using a similar strategy [110,111]. The *E. coli* platform has shown expression of a wide variety of membrane proteins including pores, channels, transporters, receptors, enzymes, and others while utilizing the exogenous addition of synthetic liposomes [106,112]

3.2.7. Difficult to Synthesize Proteins

The advantages of cell-free protein synthesis over in vivo protein synthesis, such as the open reaction and absence of living cells, allow for the production of proteins that would be difficult to manufacture in vivo due to the burden on the cell and inability to manipulate the environment of protein production (Figure 1). Such examples include antibodies, large proteins, ice structuring proteins, and metalloproteins.

Other applications, such as expression of proteins from high GC content templates (Section 3.3.2, titled "Streptomyces" and Section 3.3.7. titled "*Pseudomonas putida*") and thermostable proteins (Section 3.3.9, titled "Archaeal"), will be discussed in the low adoption section.

3.2.7.1. Antibodies

The production of functional antibodies and antibody fragments in vitro using CFPS has the potential to allow for simplification of the antibody production process for more rapid manufacturing. This advantage is due in part to the open system, which can easily be modified from case to case for the production of active antibodies using rapid design–build–test cycles and modification of the redox potential of the reaction. Antibody production has taken place in rabbit reticulocyte, *E. coli*, Chinese hamster ovary, wheat germ, and insect platforms [100,113].

One of the first instances of antibody production in a CFPS platform was the synthesis of the light chain of mouse Ig in rabbit reticulocyte [114]. Later on, the rabbit reticulocyte platform was also used to synthesize the scFv-toxin fusion protein, which contains both single-chain and gamma globulin antibodies [115,116].

Previous studies in *E. coli* have shown that protein disulfide isomerase for disulfide bond shuffling is important for active antibody formation, while addition of DsbA, a thiol disulfide oxidoreductase, does not improve active yield. This study also found that the addition of chaperones helped to increase soluble yields but not functional yield [117]. Moreover, cell-free expression has been used to overcome low yields that occur in vivo with rearrangement of variable regions [118]. In *E. coli*, synthesis of full-length correctly folded and assembled antibodies has been accomplished in a range of scales. Fab antibodies have been produced with 250 µg/mL yields in reaction scales from 60 µL to 4 L, and scFv antibodies with yields up to 800 µg/mL in reaction scales from 10 µL to 5 L. CFPS reactions containing iodoacetamide, protein disulfide isomerase, and both oxidized and reduced glutathione are used to increase active yields. These yields were also improved for industrial production by codon optimization, translation initiation optimization, and temporal assembly optimization. This demonstrates the power of CFPS for antibody production in industry as well as in screening and optimization [2]. The *E. coli* platform has also allowed for the synthesis of IgG antibody drug conjugates using genetic code expansion and iodoacetamide-treated extract supplemented with glutathione [119]. Other antibodies including the Fab fragment of 6D9, scFv to Erb-2, and even gram per liter IgG yields have been obtained in *E. coli* [120–122].

The Chinese hamster ovary platform has recently emerged as an easily optimizable platform for high yield synthesis of monoclonal antibodies (mAbs). Using a commercially available extract, successful synthesis of aglycosylated, active mAbs in yields greater than 100 µg/mL has been accomplished. The process has been taken a step further by exploring the utility of the platform for ranking yields of candidate antibodies [103]. Antibody production has also been achieved in wheat germ by lowering the concentration of DTT in the reaction or by adding protein disulfide isomerase and oxidized and reduced glutathione [123].

In the insect platform, which contains its own microsomes, adjustment of the redox potential in the reaction by omitting DTT and including glutathione allowed for the creation of antibody-enriched vesicles containing functional antibodies. This technique is notable as it mimics synthesis of antibodies as it would occur in living cells and allows for the vesicles and antibodies to be easily and efficiently separated from the CFPS reaction [124]. Moreover, single-chain antibody fragments with non-standard amino acid incorporation have been produced in the insect platform via translocation to microsomes [125]. Protein disulfide isomerase has also been supplemented to these reactions to yield more active antibodies [62].

3.2.7.2. Large Proteins

The CFPS platform makes the synthesis of very large proteins more tractable in batch mode, allowing for high quantity expression that would normally overwhelm in vivo expression methods [6]. Successful synthesis of soluble, active proteins above 100 kDa has been achieved within the *E. coli*, HeLa, insect, and rabbit reticulocyte platforms. With the high protein producing efficiency of the *E. coli* platform (Figure 2), successful synthesis of the first two (GrsA and GrsB1) of the five modules of a non-ribosomal peptide synthase (NRPS) system was completed, both of which are greater than 120 kDa in size. Specifically, these large proteins were synthesized in batch reactions that ran for 20 h and generated yields of full-length, soluble GrsA at ~106 µg/mL and GrsB1 at ~77 µg/mL [126]. HeLa cell-based CFPS platforms have also demonstrated the ability to synthesize large proteins ranging from 160 to 260 kDa. Namely, this platform produced the proteins GCN2 (160 kDa), Dicer (200 kDa), and mTOR (260 kDa) that were functionally validated with the appropriate biochemical assays [127].

B-galactosidase (116 kDa) was successfully synthesized within an insect platform [18]. The rabbit reticulocyte platform has proved to successfully synthesize active kDa proteins >100, such as a cystic fibrosis transmembrane conductance regulator of ~160 kDa [128].

3.2.7.3. Ice Structuring Proteins

Ice structuring proteins, or antifreeze proteins, are more niche, but still difficult-to-synthesize proteins that benefit greatly from CFPS. These proteins lack common structural features as a family, are difficult to express in whole cells, and require validation of protein products to ensure the active form is successfully produced. CFPS offers more rapid screening and production of both natural and engineered active ice structuring proteins. Ice structuring proteins have been produced successfully in both insect and *E. coli* platforms, and their activity can be tested without the need for purification through an ice recrystallization inhibition assay [129].

3.2.7.4. Metalloproteins

Metalloproteins, such as [FeFe] hydrogenases and multicopper oxidases (MCOs), are difficult to produce in vivo due to low yields, insolubility, poor metal cofactor assembly, and oxygen sensitivity [130,131]. However, they have the potential to enable renewable hydrogen fuel and other important biotechnological advancements. CFPS in the *E. coli* platform has enabled the manipulation of reaction conditions with chemical additives for the synthesis of soluble, active metalloproteins. Specifically, the use of post-CFPS CuSO4 addition for MCO production and the addition of maturation enzymes, iron, and sulfur for [FeFe] reductases greatly improved active enzyme yields [130,131]. Additionally, anaerobic growth of the extract source culture and anaerobic extract preparation were necessary to produce active [FeFe] reductases [131]. The H-cluster of [FeFe] hydrogenase has also been synthesized in *E. coli* CFPS through recreation of the biosynthetic pathway and used to convert apo [FeFe] hydrogenase to active protein [132].

3.2.8. Miscellaneous Applications

CFPS has also been used for a number of miscellaneous applications, including studies of translational machinery, genetic circuits, metabolic engineering, and genetic code expansion. Many of these applications are more feasible and can be used more rapidly in cell-free platforms than in vivo due to the open nature of the system, allowing for faster design–build–test cycles and direct manipulation of the reaction (Figure 1).

3.2.8.1. Studies of Translational Machinery

The open nature of CFPS and the lack of dependence on living cells enables the user to study translational machinery in ways not possible in vivo. These include ribosomal labeling, mutation of ribosomes, removal or replacement of some tRNAs, and generation of orthogonal translation systems, which can improve our understanding of the process of translation across species and help to enable a wider variety of genetic code expansion options [6,133,134]. One such study piloted hybrid ribosome platforms, by supplementing rabbit reticulocyte lysate with other mammalian ribosomes, to prevent energy consumption not directed toward protein synthesis and to boost overall yields [135]. Another study synthesized fully functional ribosomes via the integrated synthesis, assembly, and translation (iSAT) platform [136]. This was achieved through in vitro rRNA synthesis and assembly of ribosomes with supplemented *E. coli* ribosomal proteins. Functionality of these ribosomes was demonstrated by the synthesis of active protein within a single CFPS reaction [136].

3.2.8.2. Genetic Circuits

The challenge for researchers to understand the complexity of gene elements and their interplay in an expedient manner is an ongoing task. Using CFPS for modeling such genetic circuits to further understanding of the dynamics of genetic elements and to program cells capable of executing logical functions provides numerous advantages over in vivo approaches. These include (1) the control of gene and polymerase concentrations, (2) quantitative and rapid reporter measurements, and (3) a larger parameter space that can be evaluated in a high-throughput fashion [137,138]. *E. coli*, wheat germ, and yeast platforms have all exhibited utility in modeling genetic circuits, with *E. coli* extracts being the most widely used. Specifically, *E. coli* and wheat germ extracts have both modeled one-, two-, and three-stage expression cascades within a genetic circuit assembly [139]. *E coli* and yeast extracts have been used as genetic circuits to study the translational noise within cells, determine kinetic parameters, and yield insights within the construction of synthetic genetic networks [140]. Other *E. coli* genetic circuit studies have confirmed and isolated cross talking events, derived a coarse-grained enzymatic description of biosynthesis and degradation, and revealed the importance of a global mRNA turnover rate and passive competition-induced transcriptional regulation among many other studies [141–147].

3.2.8.3. Metabolic Engineering

The industrial demand for rapid development and screening of commodity chemicals and natural products has prompted the adaptation of CFPS platforms for cell-free metabolic engineering (CFME). This approach allows for a cost-effective platform to produce large amounts of diverse products in a short amount of time [148]. Specifically, CFME provides an in vitro platform comprised of catalytic proteins expressed as purified enzymes or crude lysates that are capable of being mixed to recapitulate full metabolic pathways [148]. The swift prototyping of this approach has already been employed to generate a number of diverse products using yeast and *E.coli*-based platforms [149].

The power of this approach has been used for the production of bio-ethanol using a yeast-based platform to circumvent the limitations of the conventional fermentation process. By employing a bead-beating method to generate yeast cell extract, the CFPS platform was able to generate 3.37 g/L of bio-ethanol compared to 4.46 g/L from the fermentation process at 30 °C. However, the CFPS platform excelled over the fermentation platform at higher temperatures [150]. *E. coli*-based CFME has been optimized for the metabolic conversion of glucose to 2,3-butanediol (2,3-BD) through the engineering of an *E. coli*-based extract to (1) express the genes necessary to convert pyruvate to 2,3-BD, (2) activate cell-free metabolism from glucose, and (3) optimize substrate conditions for highly productive cell-free bioconversions [151]. Additionally, *E. coli* CFME has successfully produced a high titer of mevalonate through systematic production of the enzymes involved in the mevalonate enzymatic pathway and combinatorial mixing of the lysates along with the necessary substrates to recapitulate the full mevalonate enzyme pathway in a biosynthetic manner [148]. Lastly, large NRPS proteins produced in *E. coli* CFPS underwent identical crude lysate mixing approaches to validate their functionality in a metabolic pathway and successfully produced a diketopiperazine in a 12 mg/L concentration [126].

3.2.8.4. Genetic Code Expansion

Genetic code expansion allows for site-specific incorporation of non-standard amino acids (nsAAs) into the protein of interest through reassignment of a codon. This is most commonly achieved through stop codon suppression but can also be done through sense codon reassignment, frameshift codons, or tRNA misacylation [152]. Co-production of an orthogonal tRNA in CFPS has also allowed for nsAA incorporation [153]. Applications of genetic code expansion include incorporation of biophysical probes for structural analysis by NMR, MS, and more, incorporation of fluorophores for interrogation of local protein structures, protein conjugation for production of biomaterials or protein immobilization, incorporation of post-translational modifications, and usage of photocaged amino acids for control of protein activity [152]. While genetic code expansion is possible in vivo, it requires high concentrations of often expensive nsAAs in order to increase the intracellular concentrations to levels high enough for faithful incorporation. The elimination of the cellular barrier in CFPS allows much lower concentrations of nsAA to be used, which can drastically reduce costs (Figure 1) [6,152].

Cell-free genetic code expansion has been accomplished in *E. coli,* insect, rabbit reticulocyte, and wheat germ platforms. The most extensive variety of nsAA incorporations, from hydroxytryptophan to glycosylated serine, has been achieved in *E. coli* [152]. The *E. coli* genome has even been recoded to lack the RF1 gene, and was then capable of 40 incorporations of p-acetyl phenylalanine into an elastin-like polypeptide with 98% accuracy and a 96 μg/mL yield or a single incorporation into GFP with a yield of 550 μg/mL [31,154]. Moreover, suppression of two different stop codons, enabling the incorporation of two different nsAAs into a single protein was achieved in vitro in this platform [155]. One-pot protein immobilization reactions have also been constructed in *E. coli* CFPS reactions, and are achieved using a combination of metal coordination, covalent interactions, or copper-free click chemistry between the protein and activated agarose, glass slides, beads, or silica nanoparticles [156]. This platform has even been used for screening of new aminoacyl tRNA synthetases with adjusted substrate specificity to improve incorporation of new nsAAs [119]. Furthermore, methylated oligonucleotides were utilized to sequester tRNAs in active cell extract, allowing for sense codon reassignment directly in the CFPS reaction. The oligo targets a sequence located between the anticodon and variable loop of the tRNA, and is both generic for tRNA type and species, allowing for one-pot sense codon reassignment in multiple cell-free platforms [157]. Additionally, reactions utilizing the expanded genetic code have been prepared by adding purified aminoacyl-tRNA synthetases and an orthogonal-tRNA template directly to the reaction to prevent the need for unique extract preparations for different nsAA incorporations [158].

CFPS allows for rapid screening of nsAA incorporation sites that can affect proper protein folding and yields. Insect CFPS has been used to incorporate p-azido phenylalanine, which was subsequently labeled with a fluorophore, for rapid screening of candidate incorporation sites [152,159,160]. A variety of other nsAAs have also been incorporated in insect, rabbit reticulocyte, and wheat germ platforms. A more in-depth list of many nsAAs that have been incorporated in each platform can be found in "Cotranslational Incorporation of Non-Standard Amino Acids using Cell-Free Protein Synthesis" [152].

3.3. Low Adoption Platforms

Cell-free platforms that have experienced low adoption thus far include those derived from *Neurospora crassa, Streptomyces, Vibrio natriegens, Bacillus subtilis,* tobacco, *Arabidopsis, Pseudomonas putida, Bacillus megaterium,* Archaea, and *Leishmania tarentolae* (Figure 5). These platforms were characterized as low adoption platforms because less than 25 papers have been published for each (Figure 5B and Supplementry Materials). This section will cover both platforms that were created years ago but have only been used for specialized or limited applications, newly emerging platforms, and platforms that are experiencing a revival after years with minimal usage. These platforms are generally less well optimized and well-understood than those covered in the high adoption section, but may still be of interest for certain applications or for further development. We have organized the following based upon platform rather than application to give the reader an overview of the landscape of applications that have been achieved in each platform. For platforms that have not yet had published applications, proposed applications are discussed.

3.3.1. Neurospora crassa

A platform utilizing *Neurospora crassa* was created with interest in developing it as a platform for which many gene deletion mutants exist [161]. This was proposed as a way to better study translational quality control utilizing the mutant strains available. This platform has been used to characterize the importance of 7-methylguanosine caps, determine locations of ribosome binding sites, investigate the importance of heat shocking cell cultures and prepared mRNA templates, determine kinetics of luciferase synthesis, and incorporate fluorescent nsAAs to investigate ribosomal stalling [162–167].

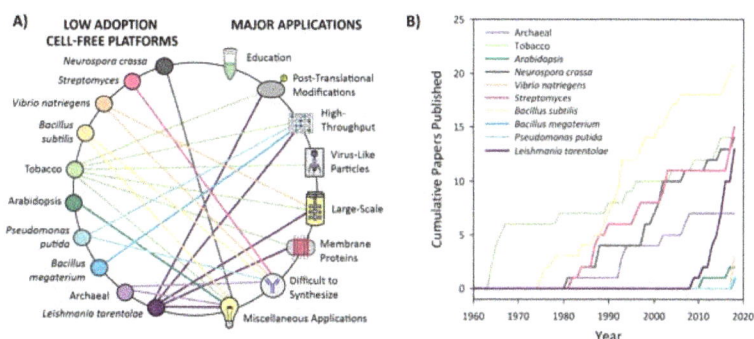

Figure 5. Low adoption cell-free platforms and their applications. (**A**) Web of the applications enabled by low adoption cell-free platforms. Connections shown are based on applications that have been published or that have been proposed in publications. Applications under "difficult to synthesize proteins" include high GC content proteins, antimicrobial peptides, pharmaceutical proteins, and thermophilic proteins. Miscellaneous applications include studies of translational machinery, investigation of antibiotic resistance, genetic circuits, metabolic engineering, and genetic code expansion. (**B**) Cumulative number of peer-reviewed publications over the last 60 years for low adoption platforms. We have used the metric of cumulative publications to indicate which platforms are less utilized and have categorized platforms with under 25 papers as low adoption platforms. These data were generated by totaling papers from a PubMed Boolean search of the following: ("cell free protein synthesis" OR "in vitro transcription translation" OR "in vitro protein synthesis" OR "cell free protein expression" OR "tx tl" OR "cell-free translation") AND "platform name." The platform name used for each search corresponds to the name listed in the graph's key. While this metric is an indicator of the level of adoption for each platform, it does suffer from inconsistencies due to irrelevant search results, such as papers reporting studies in which the researchers produce proteins from the organism of interest rather than from cell extract derived from the organism. This inconsistency was significant for platforms with fewer papers, so we pursued data curation to remove irrelevant papers and add in missing papers. This information was collected on 23 December 2018, and curated search results for each platform can be found in Supplemental Information, where red indicates that the paper was removed from the search results and green indicates that the paper was added to the search results.

3.3.2. *Streptomyces*

Streptomyces was first used in the 1980s for coupled reactions to express proteins from both linear and circular recombinant *Streptomyces* plasmids, but the original platform fell out of use in the 1990s, likely due to the time-consuming preparation and low yields of the platform [168,169]. Recently, the *Streptomyces* platform has been revived with simplified extract preparation and some improvements to protein yield [15,168]. The platform was optimized with the intention of use for expressing high GC content templates to enable production of natural gene clusters in vitro. With new genome mining technologies, knowledge of natural product gene clusters is increasing rapidly. However, in vivo expression of these clusters results in very low soluble yields due to the high metabolic burden on cells [168]. *Streptomyces*-based CFPS not only accounts for codon optimization for higher GC content templates, but also presents an opportunity for improving soluble expression of natural product gene clusters [15,168]. Examples of high GC content gene expression include *tbrP*, *tbrQ*, and *tbrN* for nonribosomal peptides synthesis of tambromycin as well as the TEII gene involved in valinomycin synthesis [168]. While the *Streptomyces* platform does significantly improve solubility of these proteins compared to expression in *E. coli* CFPS, it does suffer from diminished yields overall, indicating that further optimization of the platform is necessary [168].

3.3.3. *Vibrio natriegens*

Within the last year, the Jewett, Church, and Siemann-Herzberg laboratories have each separately developed a CFPS platform based upon *Vibrio natriegens* [10,170,171]. With its doubling time being the shortest of all known organisms, its high rate of protein synthesis, and high metabolic efficiency, this platform has potential to be an ideal candidate for CFPS [170]. In addition to its unique doubling time, *Vibrio natriegens* extract preparation requires a stationary phase harvest for the highest translational efficiency in a CFPS platform. Typically, CFPS extracts are harvested in a tight window during the mid-exponential phase to maximize translational efficiency. However, the *Vibrio natriegens* extract allows a great amount of flexibility for the user to "set and forget" the culture for a stationary phase harvest where ribosome production is thought to be lowest among other microorganisms [10].

Another advantage to extract preparation for this platform is its high resistance to damage via over-lysis. Additionally, it is relatively agnostic to lysis buffer resuspension volume. Together, these allow for inexperienced CFPS users to easily generate robust extract [10]. In addition, the *V. natriegens* platform generates a very high volume of extract compared to the standard *E. coli* platform, allowing for 8–12 mL of active lysate per L of culture compared to just 1–3 mL/L for *E. coli* when grown in shake flasks and lysed by sonication [10]. *V. natriegens* extract has even been shown to maintain 100% of activity after one week of storage at room temperature post-lyophilization in the presence of trehalose [10]. Although this platform appears to be promising in terms of flexibility and scale of extract preparation, very few applications have been proposed. Aside from reporter proteins being expressed, the Jewett laboratory has demonstrated the successful synthesis of a series of antimicrobial peptides using this platform [10].

3.3.4. *Bacillus subtilis*

The development of a *Bacillus subtilis* CFPS platform has not been pursued until recently due to requirements of exogenous mRNA addition, protease inhibitors, DNase treatments, and less efficient energy systems, as determined by studies in the 1970s and 1980s. These early studies utilized *B. subtilis* extracts to study various antibiotic resistances, investigate bacterial ribosome and mRNA specificity, and identify plasmid replication control genes [172–174]. In the last few years, the Freemont laboratory has developed a standardized workflow that circumvents the limitations of the past *B. subtilis* platform. By using a 3-phosphoglycerate (3-PGA) energy regeneration system, with optimized magnesium and potassium glutamate concentrations based upon the *E. coli* CFPS platform, the Freemont laboratory has created a *Bacillus* WB800N platform capable of expressing 0.8 µM GFPmut3b in a reaction that can last for several hours. More research is needed on this platform to reach expression levels seen within the *E. coli* platform, but the Freemont laboratory has successfully characterized an inducible expression platform that was able to quantify the activity of Renilla luciferase. Proposed applications for this platform include the production of industrial or pharmaceutical proteins and applications in metabolic engineering [16].

3.3.5. Tobacco

Though a relatively undeveloped platform, tobacco does allow for a few specific applications and is one of the few plant-based platforms. In the past decades, various parts of the tobacco plant, such as leaves, terminal buds, and trichomes, have been used to prepare extract [175–178]. These extracts were then used to elucidate differences between 70S and 80S ribosomes, understand synthesis of indoleacetic acid, diterpene cis-abienol, and cytokinins, study cauliflower mosaic virus transcription, and determine nicotine *N*-demethylase activity [176–181]. More recently, tobacco BY-2 cells have emerged as the source of extract. Preparation of up to 100 mL of cell extract from BY-2 suspension cultures is possible for larger scale applications [12]. Moreover, successful tobacco extract preparation requires only 4–5 h, whereas other eukaryotic platforms range from 1–5 days (Table 2) [64]. The BY-2 platform has enabled further investigation into positive strand RNA genomes from plant viruses, through synthesis of tomato bushy stunt virus, tomato mosaic virus, brome mosaic virus, and turnip crinkle virus [182,183]. Replicases formed from viral RNAs in CFPS are able to bind to the microsomal structures contained in the extract, allowing for elucidation of the mechanism of genome replication by these viruses, and for the screening of viral mutations [182,183].

Tobacco extract also enables some post-translational modifications, disulfide bond formation, and membrane protein synthesis. The production of a full size, active glucose oxidase antibody and a transmembrane protein has been achieved in this platform without microsomal addition, showing that the extract does contains active endogenous microsomal units that allow for disulfide bond formation, glycosylation, and co-translational membrane integration [12]. However, the full extent of possible PTMs in tobacco CFPS is not well understood. High-throughput coupled reactions from PCR templates with phosphorothioate-modified oligonucleotides have also been created with tobacco extract [12].

3.3.6. *Arabidopsis*

An *Arabidopsis*-based platform was created in 2011, with the proposed advantage of applying the vast knowledge of *Arabidopsis* genetics in combination with CFPS to study post-transcriptional regulation [184]. However, this platform has seen limited actualization of applications, with brief work done on the degradation of uncapped mRNA in mutant cell lines and some investigation into ribosome stalling [184].

3.3.7. *Pseudomonas putida*

Serving as a model organism and understood well at the biochemical level, the Gram-negative bacterium *Pseudomonas putida* has been well established for laboratory research and industrial production of biofuels, recombinant antibody fragments, and natural products. With this already well-established research framework at hand, the Jewett Laboratory has developed and optimized the *P. putida* CFPS platform capable of synthesizing approximately 200 µg/mL of reporter protein within a 4 h, 15 µL batch reaction. Extract preparation for this platform was previously reported, based on that of the *E. coli* platform with slight modifications. Overall, preparation of *P. putida* is faster and less laborious than the *Streptomyces* platform, and is hypothesized to be useful for prototyping the expression of GC-rich genes with codon usage bias. As another high GC bacteria, *P. putida* may be chosen over the *Streptomyces* platform for its aforementioned ease of extract preparation. Moving forward, this platform may also prove useful in screening gene regulatory elements, as well as closing the gap between in vitro and in vivo prediction [14].

3.3.8. *Bacillus megaterium*

In addition to the *Bacillus subtilis* platform, the Freemont laboratory has also begun to pilot a CFPS platform for *Bacillus megaterium*, a large Gram-positive bacterium with potential biotechnology applications including the production of penicillin G amidase, B-amylases, and vitamin B12. Unlike the well characterized *Bacillus subtilis* bacterium species, *B. megaterium* has remained a relatively uncharacterized microbe due to its low-efficiency and time-consuming protoplast transformation procedure. However, creating a CFPS platform to study *B. megaterium* provides some major advantages over *B. subtilis* due to its (1) stable plasmid maintenance, (2) minimal neutral alkaline protease activity, and (3) ability to metabolize low-cost substrates. Currently, this CFPS platform has been used to prototype genetic elements and has demonstrated a protein titer of about 70 µg/mL [185] (Figure 2).

3.3.9. Archaeal

Various archaeal hyperthermophiles and methanogens have been utilized to generate new CFPS platforms, including Methanobacterium formicicium, Methanosarcina barkeri, Methanococcus vannielii, Thermus Thermophilus, Sulfolobus tokodaii, Sulfolobus solfataricus, and Thermococcus kodakarensis.

The thermophilic organisms *S. solfataricus* and *T. kodakaraensis* have been utilized in CFPS for expression of thermophilic proteins, which can be difficult to synthesize in vivo. Ribosomes isolated in cell extracts from these strains are capable of performing at higher temperatures, allowing CFPS reactions to be run at higher temperatures (75 °C for *S. solfataricus*; 65 °C for *T. kodakaraensis*) for improved folding of thermophilic proteins [20,186]. However, other problems with high-temperature CFPS reactions have yet to be fully mitigated. For example, production of chitinase in *T. kodakarensis* CFPS stopped after 30 min, which was conjectured to be an issue with energy depletion worsened by the shorter half-life of energy-rich molecules at high temperatures [20]. Additionally, coupled reactions are not yet feasible at elevated temperatures, due to the differences in optimal performance temperatures for transcription and translation reactions [20].

Many archaeal methanogenic CFPS platforms have also been used to probe antibiotic sensitivity in order to elucidate phylogenetic connections. Antibiotic targeting to ribosomes can be confirmed using CFPS platforms in a way not possible in vivo because cell viability is inconsequential [187]. Antibiotic enhancement of neomycin and paromomycin and the physiological roles of polyamines were also investigated in *T. thermophilus* and *S. tokodaii* CFPS platforms [188,189].

3.3.10. *Leishmania tarentolae*

Leishmania tarentolae, a protozoan platform, is a relatively new platform that has experienced some recent optimization. *L. tarentolae* appears to be particularly promising for growth and extract scalability, with a relatively short doubling time and faster extract preparation when compared to eukaryotes of interest [190].

L. tarentolae-based CFPS has been utilized for a variety of high-throughput applications, with CFPS possible directly from PCR templates and protein analysis possible directly in the reaction mixture. One type of analysis utilizes fluorescence cross-correlation spectroscopy to analyze protein–protein or protein–small-molecule interactions [11,190]. Protein arrays can also be constructed in time and cost-effective ways in the *L. tarentolae* platform by utilizing "translation and immobilization of protein on hydrophobic substrate" (TIPoHS). Here, CFPS reactions are run on membranes, and immobilization and detection are achieved via a c-terminal GFP tag [191].

The *L. tarentolae* platform has been used for disulfide bond formation, and while other PTMs may be possible, they are not yet well defined or understood [64,192]. The platform is also capable of membrane protein synthesis with the addition of liposomes or nanodiscs, and was used to synthesize 22 different human solute carrier proteins [193]. Along with *E. coli*, methylated oligonucleotides have been used to sequester tRNAs for one-pot sense codon reassignment, allowing for genetic code expansion in *L. tarentolae* [157].

3.4. Recent and Future Applications

An incredible diversification of CFPS usage has occurred since its inception in 1961. In the last three years alone, there have been a handful of key new applications that have contributed greatly to the field of CFPS. These include the first instances of CFPS used for education, for the development of one-pot reactions for glycoprotein synthesis, for sense codon reassignment, for protein immobilization, for continued refinement of lyophilization for better shelf stability of cell-free extract, and for the demonstration of multiple non-standard amino acid incorporations into a single protein [43,46,47,53, 54,154–157]. Furthermore, a handful of promising new and revived CFPS platforms from *Streptomyces, Pseudomonas putida, Bacillus subtilis, Bacillus megaterium*, and *Vibrio natriegens* have been introduced for novel applications, including the synthesis of proteins from high GC templates (*Streptomyces*; *P. putida*), and for the further development of applications such as metabolic engineering (*B. subtilis*) [10,14–16, 168,170,171].

Despite the proliferation of CFPS platforms and applications in the last 60 years, there are still many new directions in which the technology can be taken. Some future directions for CFPS may include further development and optimization of current platforms, especially the emerging or re-emerging platforms of *Streptomyces, Pseudomonas, Bacillus subtilis, Bacillus megaterium*, and *Vibrio natriegens*. Soon, the proposed applications of these platforms may be actualized. Furthermore, we may see additional CFPS platforms be established to solve new problems or to fill other existing gaps that current platforms have left. In terms of applications, there may be more utilization of CFPS for education, metabolic engineering, personalized medicine, and diagnostics, as current work seems to have only scratched the surface of these applications. Further development of large-scale CFPS may also be a future direction developed alongside these applications in order to support new industrial endeavors.

4. Methodological Differences between Platforms

While the user's selection of a given CFPS platform will be primarily driven by the applications enabled by a platform, there are often multiple platforms that can be used for a single type of application. The choice between these platforms can be guided by factors including the accessibility and technical complexity of the methods used to produce the cell extract, the reagents used for CFPS reactions, the type of reaction (coupled vs uncoupled), and the productivity of the platform. Here we provide further guidance to the user in choosing the platform that best suits their needs, and simplify the effort needed to make this choice by providing a condensed methodological comparison of the high adoption cell-free platforms: *E. coli*, insect, yeast, Chinese hamster ovary, rabbit reticulocyte lysate, wheat germ, and HeLa cells (Tables 1–3).

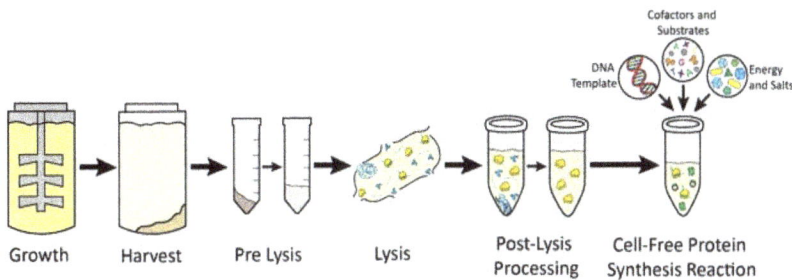

Figure 6. General workflow for preparation of cell-free extract and set up of CFPS reactions. A visualization from cell growth to the CFPS reaction is depicted above for a new user, highlighting the main steps involved.

4.1. Productivity

Firstly, different platforms will be better suited for the production of different proteins of interest, and maximizing protein yields is not required for all applications. Therefore, matching the application with a platform's productivity will enrich success for new users (Figure 2). For example, industrial level protein production is currently best enabled by *E. coli* or wheat germ platforms, with possibilities of large-scale protein production in the emerging *Vibrio natriegens* and *Pseudomonas putida* platforms (Figures 4 and 5). However, for many applications that may not require large protein samples, such as small-scale assays or functional investigations, most possible platforms would still provide large enough yields. In general, eukaryotic platforms give lower protein yields, with the exception of the wheat germ platform (Figure 2). On the lowest end of the productivity scale are the rabbit reticulocyte and archaeal platforms, which produce under 20 μg/mL of protein in batch format (Figure 2). Overall, it is important to choose a platform that is suited to producing the protein of interest in the quantity necessary for the desired application.

4.2. Growth

Methodology for cell growth from representative sources for each high adoption platform is summarized in Table 1. Growth media is highly variable between platforms, as would be expected even in in vivo protein expression. Additionally, cells can be grown in a variety of vessels, from baffled flasks in an incubator for wheat germ and *E. coli* to fermenters and spinner flasks for insect, Chinese hamster ovary, and HeLa cells. The vessel choice may also depend on the growth scale desired. Lastly, cell cultures must be harvested, which is typically done via centrifugation and washing of the pelleted cells. Platforms that stand out most due to specialized methods are wheat germ and rabbit reticulocyte. In general, all other platforms utilize cell growth in liquid culture, centrifugation for the harvest of cells, and pellet washing in an HEPES-based buffer supplemented with acetate salts and with DTT in some cases. However, for wheat germ, wheat seeds are ground in a mill and sieved, and embryos are selected by solvent flotation [194]. Rabbit reticulocyte extract preparation may even require treatment of live rabbits to make them anemic as well as bleeding of the rabbits to obtain the cells needed [98].

Table 1. Comparison of growth methods for high adoption platforms. We have assembled the major growth methodologies for each of the high adoption platforms to give users an idea of the relative differences between them. These are not the only techniques that have been used for growth for each platform, but they are generally representative of the methods.

Platform	Growth		Key Citations
	Media/Vessel	Harvest	
E. coli	**Media:** 2× YPTG (5 g NaCl, 16 g Tryptone, 10 g Yeast extract, 7 g KH$_2$PO$_4$, 3 g KHPO$_4$, pH 7.2/750 mL solution, 18 g Glucose/250 mL solution). **Vessel:** 2 L Baffled Flask. **Conditions:** 37 °C, 200 RPM	When OD$_{600}$ is 3, centrifuge at 5000× *g* for 10 min at 10 °C. Wash pellet with 30 mL S30 buffer (10 mM Tris OAc, pH 8.2, 14 mM Mg(OAc)$_2$, 60 mM KOAc, 2 mM DTT), then centrifuge at 5000× *g* for 10 min at 10 °C. Repeat wash 3 times in total.	[22]
Wheat Germ	Grind wheat seeds in a mill.	Sieve through 710–850 mm mesh, select embryos via solvent flotation method using a solvent containing 240:600 *v/v* cyclohexane and carbon tetrachloride. Dry in fume hood overnight.	[194]
Yeast	**Media:** 2% *w/v* Peptone, 1% *w/v* Yeast extract, 2% *w/v* Glucose **Vessel:** 2.5 L baffled flask **Conditions:** 30 °C, 250 RPM	When OD$_{600}$ of 10–12 is reached, centrifuge culture for 10 min at 3000× *g*. Wash pellet with Buffer A (20 mM HEPES, pH 7.4, 100 mM KOAc, 2 mM Mg(OAc)$_2$). Centrifuge for 10 min at 3000× *g* and 4 °C. Repeat washing 3 times.	[19]
Rabbit Reticulocyte	Make rabbits anemic over 3 days by injections of APH.	Bleed rabbits on day 8. Filter blood through cheesecloth and keep on ice, then centrifuge at 2000 RPM for 10 min.	[98]
Insect	**Media:** Animal component free insect cell medium. **Vessel:** Fermentor. **Conditions:** 27 °C	When cell density reaches 4 × 10^6 cell/mL, centrifuge culture at 200× *g* for 10 min. Wash once with buffer (40 mM HEPES KOH, pH 7.5, 100 mM KOAC, 4 mM DTT).	[129]
HeLa	**Media:** Minimal essential medium supplemented with 10% heat-inactivated fetal calf serum, 2 mM glutamine, 1 U/mL penicillin, 0.1 mg/mL streptomycin. **Vessel:** Spinner flask with cell culture controller **Conditions:** 37 °C, pH 7.2, 67 ppm oxygen, 50 RPM	Harvest when cell density reaches 0.7–0.8 × 10^6 cells/mL. Wash 3 times with buffer (35 mM HEPES KOH, pH 7.5, 140 mM NaCl, 11 mM glucose).	[13]
Chinese Hamster Ovary	**Media:** Power Chinese hamster ovary-2 chemically defined serum-free media **Vessel:** Fermentor **Conditions:** 37 °C	Harvest at 2 × 10^6 cells/mL cell density by centrifuging at 200× *g* for 10 min. Wash cells once with buffer (40 mM HEPES KOH, pH 7.5, 100 mM NaOAc, 4 mM DTT).	[17]

4.3. Extract Preparation

Extract preparation consists of pre-lysis preparation, lysis, and post-lysis processing, which are covered in detail for each high adoption platform in Table 2. Lysis methods not only vary from platform to platform, but many different lysis methods can also be used for a single platform. Here we have highlighted just one of the methods used for each platform, but others may also be viable. Firstly, cells are resuspended, then sonication (*E. coli*, wheat germ), homogenization (yeast), nitrogen cavitation (HeLa, insect), freeze-thaw (insect), syringing (Chinese hamster ovary), osmotic lysis (rabbit reticulocyte), or other lysis methods may be used to disrupt cell membranes. The lysate is centrifuged at high speeds to separate out cell membrane fragments and other unnecessary cellular debris. Post-processing after lysis and centrifugation also varies from platform to platform. For example, a run-off reaction, where the supernatant is incubated, is performed on *E. coli* extract. For Chinese hamster ovary, HeLa, insect, wheat germ, and yeast, desalting or dialysis is performed on the supernatant. The Chinese hamster ovary, HeLa, and rabbit platforms are generally treated with micrococcal nuclease to degrade remaining endogenous mRNA in the extract, and the nuclease activity is quenched through addition of EGTA. All extracts are then flash frozen in liquid nitrogen and stored either in liquid nitrogen, or more frequently at −80 °C if CFPS is not immediately performed afterwards.

Table 2. Comparison of extract preparation methods for high adoption platforms. We have assembled the major extract preparation methodology for each of the high adoption platforms to give users an idea of the relative differences between them. These are not the only techniques that have been used for extract preparation for each platform, but they are generally representative of the methods.

	Extract Prep				
Platform	Pre-Lysis	Lysis	Post-Lysis Processing	Growth and Prep Time	Key Citations
E. coli	Resuspend in 1 mL/1 g pellet of S30 buffer by vortexing.	Sonicate on ice for 3 cycles of 45 s on, 59 s off at 50% amplitude. Deliver 800–900 J total for 1.4 mL of resuspended pellet. Supplement with a final concentration of 3 mM DTT.	Centrifuge lysate at 18,000× g and 4 °C for 10 min. Transfer supernatant while avoiding pellet. Perform runoff reaction on supernatant at 37 °C and 250 RPM for 60 min. Centrifuge at 10,000× g and 4 °C for 10 min. Flash freeze supernatant and store at −80 °C.	1–2 days	[22]
Wheat Germ	Wash 3 times with water under vigorous stirring to remove endosperm.	Sonicate for 3 min in 0.5% Nonidet P-40. Wash with sterile water. Grind washed embryos into fine powder in liquid nitrogen and resuspend 5 g in 5mL of 2× Buffer A (40 mM HEPES, pH 7.6, 100 mM KOAc, 5 mM Mg(OAc)₂, 2 mM CaCl, 4 mM DTT, 0.3 mM each of the 20 amino acids).	Centrifuge at 30,000× g for 30 min. Filter supernatant through G-25 column equilibrated with Buffer A. Centrifuge column product at 30,000× g for 10 min. Adjust to 200 A₂₆₀/mL with Buffer A. Store in liquid nitrogen.	4–5 days	[64,194]
Yeast	Resuspend cells in 1 mL lysis buffer (20 mM HEPES KOH, pH 7.4, 100 mM KOAc, 2 mM Mg(OAc)2, 2 mM DTT, 0.5 mM PMSF) per 1 g cell pellet.	Pass through homogenizer once at 30,000 psig.	Centrifuge homogenate at 30,000× g for 30 min at 4 °C. Then repeat centrifugation with supernatant in a spherical bottom bottle. Desalt supernatant in dialysis tubing with 4 exchanges of 50× volume lysis buffer for 30 min each at 4 °C. Centrifuge at 60,000× g for 20 min at 4 °C. Flash freeze and store at −80 °C.	1–2 days	[19,195]
Rabbit Reticulocyte	Resuspend cells in buffered saline with 5 mM glucose, then centrifuge at 2000 RPM for 10 min. Repeat wash 3 times.	Resuspend cells in 1.5 times the packed cell volume of ice-cold water, mix thoroughly.	Spin lysate at 15,000× g for 20 min at 2 °C. Pour supernatant though 53 μm nylon. Treat with micrococcal nuclease by adding 0.2 mL of 1 mM hemin, 0.1 mL of 5 mg/mL creatine kinase, 0.1 mL of 0.1 M CaCl₂, 0.1 mL of micrococcal nuclease. Incubate at 20 °C for 15 min, then add 0.1 mL of 0.2 M EGTA and 60 μL of 10 mg/mL tRNA. Store in liquid nitrogen or at −70 °C.	~8 days to treat rabbits, 1 day for extract preparation	[98]
Insect	Resuspend cells in buffer to final density of 2 × 10⁸ cells/mL.	Mechanically lyse cells by rapidly freezing in liquid nitrogen, then thawing in water bath at 4 °C or use a disruption chamber with 20 kg/cm² nitrogen gas for 30 min.	Centrifuge lysate at 10,000× g for 10 min. Apply supernatant to G-25 gel filtration column. Pool fractions with highest A₂₆₀, flash freeze in liquid nitrogen and store at −80 °C.	1–2 days	[18,129,195,196]
HeLa	Resuspend in extraction buffer (20 mM HEPES KOH, pH 7.5, 135 mM KOAc, 30 mM KCl, 1.655 mM Mg(OAc)₂) to ~2.3 × 10⁸ cells/mL.	Disrupt cells via 1 MPa nitrogen pressure for 30 min in a cell disruption chamber.	Centrifuge homogenate at 10,000× g for 5 min at 4 °C. Pass supernatant through a PD-10 desalting column equilibrated with extraction buffer. Treat 100 μL of extract with 1 μL of 7500 U/mL nuclease S7 and 1 μL of 100 mM CaCl₂ for 5 min at 23 °C, then add 8 μL of 30 mM EGTA. Flash freeze eluted extract in liquid nitrogen and store at −80 °C.	1–2 days	[13,195]
Chinese Hamster Ovary	Resuspend cells in buffer to density of 5 × 10⁸ cells/mL.	Disrupt cells by syringing the pellet through a 20-gauge needle.	Centrifuge lysate at 10,000× g for 10 min. Apply supernatant to G-25 gel filtration column. Pool fractions with an A₂₆₀ above 100. Treat pooled fractions with 10 U/mL S7 nuclease and 1 mM CaCl₂, incubate at room temperature for 2 min, then add 6.7 mM EGTA. Flash freeze in liquid nitrogen and store at −80 °C.	1–2 days	[17]

4.4. CFPS Reaction Setup

CFPS reaction setup requires mixing of many reagents to initiate protein synthesis, and the details of setup for each high adoption platform are covered in Table 3. There are two main differences among CFPS setups: the chosen energy system and whether the reaction is coupled or uncoupled. Otherwise, the reaction components are generally the same, with two unique reagents used for each platform and slight variations in concentration from platform to platform. Common reagents include ATP, GTP, UTP,

CTP, tRNA, HEPES, Mg salts, K salts, 20 amino acids, and energy rich molecules. Most platforms use a creatine phosphate/creatine kinase energy system, and the most work has been done in *E. coli* to enable more inexpensive energy systems, such as PEP, glucose, and maltodextrin [8,197]. Reaction temperature has also been a major point of optimization for each of these platforms, with typical temperatures ranging from 21 to 37 °C among the various platforms [17,195,198] (Table 2). In terms of reaction type, coupled reactions are desirable because of the ease of setup, but uncoupled reactions are typically used for eukaryotic platforms to improve yields (see Section 2.2, titled "Coupled and Uncoupled Formats") [28]. Uncoupled reactions require an in vitro transcription reaction often catalyzed by T7 RNA polymerase (T7RNAP), followed by mRNA purification, then a cell-free translation reaction utilizing the prepared lysate, and are both more time-consuming and more difficult in terms of handling. Platforms that generally utilize uncoupled reactions include wheat germ, rabbit reticulocyte, insect, and HeLa. Transcription for most platforms that utilize coupled reactions require T7RNAP, but some platforms, such as *E. coli* are able to employ solely the endogenous polymerase [199,200].

Table 3. Comparison of cell-free protein synthesis reaction setup for high adoption platforms. This table is intended to help users understand major differences between setups for various high adoption platforms, namely whether reactions are generally coupled or uncoupled, what energy systems are typical, and what temperatures the reactions are run at. These are not the only setups that have been used for successful cell-free protein expression in each platform, but they are generally representative of the reagents, concentrations, and temperatures used for each platform.

Platform	Vessel/Conditions	Cell-Free Protein Synthesis Reaction Reaction Composition	Energy Systems	Key Citations
E. coli	**Vessel:** Many vessels can be used, yield increases as the surface area to reaction volume ratio increases **Conditions:** 30 °C overnight or 37 °C for 4 h	33% v/v *E. coli* extract, 16 µg/mL T7RNAP, 16 ng/mL DNA template, Solution A (1.2 mM ATP, 0.85 mM GTP, 0.85 mM UTP, 0.85 mM CTP, 31.50 µg/mL Folinic Acid, 170.60 µg/mL tRNA, 0.40 mM Nicotinamide Adenine Dinucleotide (NAD), 0.27 mM Coenzyme A (CoA), 4 mM Oxalic Acid, 1 mM Putrescine, 1.50 mM Spermidine, 57.33 mM HEPES buffer), Solution B (10 mM Mg(Glu)$_2$, 10 mM NH$_4$(Glu), 130 mM K(Glu), 2 mM of each amino acid, 0.03 M Phosphoenolpyruvate (PEP))	PEP, glucose + glutamate decarboxylase, or maltodextrin are possible	[22,201]
Wheat Germ	**Vessel:** Not noted **Conditions:** 26 °C	First, perform an in vitro transcription reaction and isolate mRNA using SP6 RNA polymerase. Set up cell-free translation as follows: 24% v/v wheat germ extract, 4 mM HEPES KOH, pH 7.8, 1.2 mM ATP, 0.25 mM GTP, 16 mM creatine phosphate, 0.45 mg/mL creatine kinase, 2 mM DTT, 0.4 mM spermidine, 0.3 mM of each of the 20 amino acids, 2.5 mM Mg(OAc)$_2$, 100 mM KOAc, 50 µg/mL deacylated tRNA from wheat embryos, 0.05% Nonidet P-40, 1 µM E-64 as proteinase inhibitor, 0.005% NaN$_3$, 0.02 nmol mRNA.	Creatine phosphate + creatine kinase	[194]
Yeast	**Vessel:** 15 µL reactions in 1.5 mL microfuge tubes **Conditions:** 21 °C	25 mM HEPES KOH, pH 7.4, 120 mM K(Glu), 6 mM Mg(Glu)$_2$, 1.5 mM ATP, 2 mM GTP, 2 mM CTP, 2 mM UTP, 0.1 mM of each of 20 amino acids, 25 mM creatine phosphate, 2 mM DTT, 0.27 mg/mL creatine phosphokinase, 200 U/mL RNase Inhibitor, 27 µg/mL T7 RNAP, DNA template, and 50% v/v yeast extract	Creatine phosphate + creatine kinase	[19,195]
Rabbit Reticulocyte	**Vessel:** 200 µL reaction performed in an NMR spectrometer **Conditions:** 30 °C	First, perform an in vitro transcription reaction and isolate mRNA using T7 RNAP. Supplement 1 mL of rabbit reticulocyte lysate with 25 µM hemin, 25 µg creatine kinase, 5 mg phosphocreatine, 50 µg of bovine liver tRNAs, and 2 mM D-glucose. Initiate in vitro translation by combining 27 nM of in vitro transcribed mRNAs, 50% v/v supplemented lysate, 75 mM KCl, 0.75 mM MgCl$_2$, and 20 µM amino acids mix.	Creatine phosphate + creatine kinase	[135]
Insect	**Vessel:** 25 µL reaction, vessel size not noted **Conditions:** 25 °C	First, perform an in vitro transcription reaction and isolate mRNA using T7 RNAP. Then set up cell-free translation as follows: 1.5 mM Mg(OAc)$_2$, 0.25 mM ATP, 0.1 mM GTP, 0.1 mM EGTA, 40 mM HEPES KOH, pH 7.9, 100 mM KOAc, 20 mM creatine phosphate, 200 µg/mL creatine kinase, 2 mM DTT, 80 µM of each of the 20 amino acids, 0.5 mM PMSF, 1 U/µL RNase inhibitor, 200 µg/mL tRNA, 320 µg/mL mRNA, and 50% v/v insect cell extract. Addition of 20% v/v glycerol to the reaction was also shown to improve yields.	Creatine phosphate + creatine kinase	[18]
HeLa	**Vessel:** 6 µL reaction, vessel not noted **Conditions:** 32 °C, 1 h	First, perform an in vitro transcription reaction and isolate mRNA using T7 RNAP. Cell-free translation is performed as follows: 75% v/v HeLa cell extract, 30 µM of each of the 20 amino acids, 27 mM HEPES KOH, pH 7.5, 1.2 mM ATP, 0.12 mM GTP, 18 mM creatine phosphate, 0.3 mM spermidine, 44–224 mM KOAc, 16 mM KCl, 1.2 mM Mg(OAc)$_2$, 90 µg/mL calf liver tRNA, 60 µg/mL creatine kinase, and purified mRNA.	Creatine phosphate + creatine kinase	[13]
Chinese Hamster Ovary	**Vessel:** 25 µL reaction, vessel size not noted **Conditions:** 33 °C, 500 RPM shaking in thermomixer	25% v/v Chinese hamster ovary cell extract, 100 µM of each of the 20 amino acids, 1.75 mM ATP, 0.30 mM CTP, 0.30 mM GTP, 0.30 mM UTP, 20 nM DNA template, 1 U/µL T7 RNAP, 30 mM HEPES KOH, pH 7.6, 150 mM KOAc, 3.9 mM Mg(OAc)$_2$, 20 mM creatine phosphate, 100 µg/mL creatine kinase, 0.25 mM spermidine, and 2.5 mM DTT.	Creatine phosphate + creatine kinase	[17]

4.5. Time

Overall, wheat germ and rabbit reticulocyte are the most time-consuming preparations, at 4–5 days for wheat germ and up to 9 days for rabbit, if treatment of animals is needed. All other platforms hover around the 1–2 day mark for preparation, with highly variable growth times dependent on doubling time for the strain and final cell density desired for harvest. *E. coli* requires the least time for preparation from the initiation of culture growth to the final freezing of extract due to its quick doubling time and relatively simple extract preparation procedure.

5. Standard Optimizations

A variety of internal development of the CFPS platforms is constantly occurring in order to improve protein yields and streamline extract preparation. Some major advances have greatly improved a variety of the CFPS platforms, such as internal ribosome entry sites (IRESs), species-independent translational leaders (SITS), and 5′UTR optimization. These have improved the rates of translation in eukaryotic platforms, which can limit protein yield. 5′UTRs are used to mimic cap structures and promote binding of the ribosome to the mRNA template, but in some cases they have also been found to be unhelpful or even detrimental to productivity. Additionally, 5′UTR choice may require some testing and optimization before application [6,11,19,202]. IRESs are sequences utilized by viruses to hijack cellular machinery for replication. They have been added to CFPS templates in order to bypass translation initiation factors, but many are species-dependent. However, IRESs have been used in rabbit reticulocyte, Chinese hamster ovary, yeast, and *Leishmania tarentolae* [64,203–205]. SITS are unstructured translation leaders that allow transcribed mRNA to interact directly with ribosomes across a variety of CFPS platforms from many cell types, such that translation initiation factors are not needed [11,190,193]. Codon optimization of the template DNA has also been used to improve yields in eukaryotic platforms [73].

In addition to template optimization, many high adoption platforms have undergone optimization of cell-free reaction reagent concentrations through systematic titrations of the main reagents [197,206]. Additionally, protein yields can be augmented by the addition of purified transcriptional and translational components or molecular crowding agents [207,208].

6. Conclusions

This review is aimed at helping new users of CFPS platforms determine which platform best suits their needs. We sought to highlight similarities and differences among the platforms, the applications that can be achieved by each, and the reasons one platform may be more advantageous for a certain goal than another.

We recommend new users first investigate the high adoption platforms to find one that suits them, as these platforms have been best optimized and there is plentiful literature to support the user. High adoption platforms include *E. coli*, insect, yeast, Chinese hamster ovary, rabbit reticulocyte, wheat germ, and HeLa. For these platforms, we have covered a wide spectrum of applications that are enabled by each, to provide the reader with an idea of the breadth of possibilities in CFPS, as well as to hopefully cover a wide spectrum of user needs. These applications include education, post-translational modifications, high-throughput expression, virus-like particles production, large-scale synthesis, membrane proteins, difficult-to-synthesize proteins (antibodies, large proteins, ice-structuring proteins, and metalloproteins), and miscellaneous applications (studies of translational machinery, genetic code expansion, metabolic engineering, and genetic circuits). In addition, we have covered the methods for growth, extract preparation, and cell-free reaction setup, as well as batch reaction protein yield, such that the reader can further determine if the platform suits their needs and obtain a better understanding of what is required for successful implementation of each.

We also briefly covered the applications enabled by low adoption platforms including Neurospora crassa, Streptomyces, Vibrio natriegens, Bacillus subtilis, tobacco, Arabidopsis, Pseudomonas putida,

Bacillus megaterium, Archaea, and Leishmania tarentolae. While these platforms have some work supporting their use, they have generally been used by only a few and are not as well optimized. However, these platforms may still provide some key advantages to the field if more work is done with them. Additionally, the emerging platforms of Vibrio natriegens, Streptomyces, Bacillus subtilis, Bacillus megaterium, and Pseudomonas putida are proposed to enable exciting new applications of CFPS, including natural product synthesis from high GC templates.

Supplementary Materials: The following are available online at http://www.mdpi.com/2409-9279/2/1/24/s1: Spreadsheet 1: PubMed search results for high adoption platforms; Spreadsheet 2: Curated PubMed search results for low adoption platforms.

Funding: This research was funded by the Bill and Linda Frost Fund, Center for Applications in Biotechnology's Chevron Biotechnology Applied Research Endowment Grant, Cal Poly Research, Scholarly, and Creative Activities Grant Program (RSCA 2017), and the National Science Foundation (NSF-1708919). MZL acknowledges the California State University Graduate Grant.

Acknowledgments: The authors thank K.R. Watts, W.Y. Kao, and L.C. Williams for helpful discussions.

Conflicts of Interest: The authors declare no conflict of interest.

References

1. Nirenberg, M.W.; Matthaei, J.H. The dependence of cell-free protein synthesis in E. coli upon naturally occurring or synthetic polyribonucleotides. *Proc. Natl. Acad. Sci. USA* **1961**, *47*, 1588–1602. [CrossRef] [PubMed]
2. Yin, G.; Garces, E.D.; Yang, J.; Zhang, J.; Tran, C.; Steiner, A.R.; Roos, C.; Bajad, S.; Hudak, S.; Penta, K.; et al. Aglycosylated antibodies and antibody fragments produced in a scalable in vitro transcription-translation system. *MAbs* **2012**, *4*, 217–225. [CrossRef] [PubMed]
3. Woodrow, K.A.; Swartz, J.R. A sequential expression system for high-throughput functional genomic analysis. *Proteomics* **2007**, *7*, 3870–3879. [CrossRef] [PubMed]
4. Carlson, E.D.; Gan, R.; Hodgman, E.C.; Jewet, M.C. Cell-Free Protein Synthesis: Applications Come of Age. *Biotechnol. Adv.* **2014**, *30*, 1185–1194. [CrossRef] [PubMed]
5. Focke, P.J.; Hein, C.; Hoffmann, B.; Matulef, K.; Bernhard, F.; Dötsch, V.; Valiyaveetil, F.I. Combining in Vitro Folding with Cell Free Protein Synthesis for Membrane Protein Expression. *Biochemistry* **2016**, *55*, 4212–4219. [CrossRef]
6. Rosenblum, G.; Cooperman, B.S. Engine out of the chassis: Cell-free protein synthesis and its uses. *FEBS Lett.* **2014**, *588*, 261–268. [CrossRef] [PubMed]
7. Shrestha, P.; Holland, T.M.; Bundy, B.C. Streamlined extract preparation for Escherichia coli-based cell-free protein synthesis by sonication or bead vortex mixing. *Biotechniques* **2012**, *53*, 163–174.
8. Caschera, F.; Noireaux, V. Synthesis of 2.3 mg/ml of protein with an all Escherichia coli cell-free transcription-translation system. *Biochimie* **2014**, *99*, 162–168. [CrossRef]
9. Harbers, M. Wheat germ systems for cell-free protein expression. *FEBS Lett.* **2014**, *588*, 2762–2773. [CrossRef]
10. Des Soye, B.J.; Davidson, S.R.; Weinstock, M.T.; Gibson, D.G.; Jewett, M.C. Establishing a High-Yielding Cell-Free Protein Synthesis Platform Derived from Vibrio natriegens. *ACS Synth. Biol.* **2018**, *7*, 2245–2255. [CrossRef]
11. Mureev, S.; Kovtun, O.; Nguyen, U.T.T.; Alexandrov, K. Species-independent translational leaders facilitate cell-free expression. *Nat. Biotechnol.* **2009**, *27*, 747–752. [CrossRef]
12. Buntru, M.; Vogel, S.; Stoff, K.; Spiegel, H.; Schillberg, S. A versatile coupled cell-free transcription-translation system based on tobacco BY-2 cell lysates. *Biotechnol. Bioeng.* **2015**, *112*, 867–878. [CrossRef]
13. Mikami, S.; Masutani, M.; Sonenberg, N.; Yokoyama, S.; Imataka, H. An efficient mammalian cell-free translation system supplemented with translation factors. *Protein Expr. Purif.* **2006**, *46*, 348–357. [CrossRef]
14. Wang, H.; Li, J.; Jewett, M.C. Development of a Pseudomonas putida cell-free protein synthesis platform for rapid screening of gene regulatory elements. *Synth. Biol.* **2018**, *3*. [CrossRef]
15. Li, J.; Wang, H.; Jewett, M.C. Expanding the palette of Streptomyces-based cell-free protein synthesis systems with enhanced yields. *Biochem. Eng. J.* **2018**, *130*, 29–33. [CrossRef]

16. Kelwick, R.; Webb, A.J.; MacDonald, J.T.; Freemont, P.S. Development of a Bacillus subtilis cell-free transcription-translation system for prototyping regulatory elements. *Metab. Eng.* **2016**, *38*, 370–381. [CrossRef]

17. Brödel, A.K.; Sonnabend, A.; Kubick, S. Cell-free protein expression based on extracts from CHO cells. *Biotechnol. Bioeng.* **2014**, *111*, 25–36. [CrossRef]

18. Ezure, T.; Suzuki, T.; Higashide, S.; Shintani, E.; Endo, K.; Kobayashi, S.; Shikata, M.; Ito, M.; Tanimizu, K. Cell-Free Protein Synthesis System Prepared from Hi5 Insect Cells by Freeze-Thawing. *Biotechnol. Prog* **2006**, *22*, 1570–1577. [CrossRef]

19. Gan, R.; Jewett, M.C. A combined cell-free transcription-translation system from Saccharomyces cerevisiae for rapid and robust protein synthesis. *Biotechnol. J.* **2014**, *9*, 641–651. [CrossRef]

20. Endoh, T.; Kanai, T.; Sato, Y.T.; Liu, D.V.; Yoshikawa, K.; Atomi, H.; Imanaka, T. Cell-free protein synthesis at high temperatures using the lysate of a hyperthermophile. *J. Biotechnol.* **2006**, *126*, 186–195. [CrossRef]

21. Kobs, G. *Selecting the Cell-Free Protein Expression System That Meets Your Experimental Goals*; Promega Corporation: Madison, WI, USA, 2008.

22. Levine, M.Z.; Gregorio, N.E.; Jewett, M.C.; Watts, K.R.; Oza, J.P. Escherichia coli-based cell-free protein synthesis: Protocols for a robust, flexible, and accessible platform technology. *J. Vis. Exp.* **2019**, in press. [CrossRef]

23. Ohashi, H.; Kanamori, T.; Shimizu, Y.; Ueda, T. A Highly Controllable Reconstituted Cell-Free System -a Breakthrough in Protein Synthesis Research. *Curr. Pharm. Biotechnol.* **2010**, *11*, 267–271. [CrossRef]

24. Shimizu, Y.; Kanamori, T.; Ueda, T. Protein synthesis by pure translation systems. *Methods* **2005**, *36*, 299–304. [CrossRef]

25. Shimizu, Y.; Inoue, A.; Tomari, Y.; Suzuki, T.; Yokogawa, T.; Nishikawa, K.; Ueda, T. Cell-free translation reconstituted with purified components. *Nat. Biotechnol.* **2001**, *19*, 751–755. [CrossRef]

26. Hillebrecht, J.R.; Chong, S. A comparative study of protein synthesis in in vitro systems: From the prokaryotic reconstituted to the eukaryotic extract-based. *BMC Biotechnol.* **2008**, *8*, 1–9. [CrossRef]

27. Karikó, K.; Muramatsu, H.; Welsh, F.A.; Ludwig, J.; Kato, H.; Akira, S.; Weissman, D. Incorporation of Pseudouridine Into mRNA Yields Superior Nonimmunogenic Vector With Increased Translational Capacity and Biological Stability. *Mol. Ther.* **2008**, *16*, 1833–1840. [CrossRef]

28. Hansen, M.M.K.; Ventosa Rosquelles, M.; Yelleswarapu, M.; Maas, R.J.M.; Van Vugt-Jonker, A.J.; Heus, H.A.; Huck, W.T.S. Protein Synthesis in Coupled and Uncoupled Cell-Free Prokaryotic Gene Expression Systems. *ACS Synth. Biol.* **2016**, *5*, 1433–1440. [CrossRef]

29. Spirin, A.S. High-throughput cell-free systems for synthesis of functionally active proteins. *Trends Biotechnol.* **2004**, *22*, 538–545. [CrossRef]

30. Stech, M.; Quast, R.B.; Sachse, R.; Schulze, C.; Wüstenhagen, D.A.; Kubick, S. A Continuous-Exchange Cell-Free Protein Synthesis System Based on Extracts from Cultured Insect Cells. *PLoS ONE* **2014**, *9*, e96635. [CrossRef]

31. Hong, S.H.; Kwon, Y.-C.; Martin, R.W.; Des Soye, B.J.; de Paz, A.M.; Swonger, K.N.; Ntai, I.; Kelleher, N.L.; Jewett, M.C. Improving cell-free protein synthesis through genome engineering of Escherichia coli lacking release factor 1. *Chembiochem* **2015**, *16*, 844–853. [CrossRef]

32. Endo, Y.; Otsuzuki, S.; Ito, K.; Miura, K. Production of an enzymatic active protein using a continuous flow cell-free translation system. *J. Biotechnol.* **1992**, *25*, 221–230. [CrossRef]

33. Volyanik, E.V.; Dalley, A.; Mckay, I.A.; Leigh, I.; Williams, N.S.; Bustin, S.A. Synthesis of Preparative Amounts of Biologically Active Interleukin-6 Using a Continuous-Flow Cell-Free Translation System. *Anal. Biochem.* **1993**, *214*, 289–294. [CrossRef]

34. Thoring, L.; Kubick, S. Versatile cell-free protein synthesis systems based on chinese hamster ovary cells. In *Methods in Molecular Biology*; Humana Press: New York, NY, USA, 2018; Volume 1850, pp. 289–308. ISBN 978-1-4939-8729-0.

35. Schoborg, J.A.; Hodgman, C.E.; Anderson, M.J.; Jewett, M.C. Substrate replenishment and byproduct removal improve yeast cell-free protein synthesis. *Biotechnol. J.* **2014**, *9*, 630–640. [CrossRef]

36. Schwarz, D.; Junge, F.; Durst, F.; Frölich, N.; Schneider, B.; Reckel, S.; Sobhanifar, S.; Dötsch, V.; Bernhard, F. Preparative scale expression of membrane proteins in Escherichia coli-based continuous exchange cell-free systems. *Nat. Protoc.* **2007**, *2*, 2945–2957. [CrossRef]

37. Chekulayeva, M.N.; Kurnasov, O.V.; Shirokov, V.A.; Spirin, A.S. Continuous-Exchange Cell-Free Protein-Synthesizing System: Synthesis of HIV-1 Antigen Nef. *Biochem. Biophys. Res. Commun.* **2001**, *280*, 914–917. [CrossRef]

38. Quast, R.B.; Sonnabend, A.; Stech, M.; Wustenhagen, D.A.; Kubick, S. High-yield cell-free synthesis of human EGFR by IRES-mediated protein translation in a continuous exchange cell-free reaction format. *Sci. Rep.* **2016**, *6*, 30399. [CrossRef]

39. Thoring, L.; Dondapati, S.K.; Stech, M.; Wustenhagen, D.A.; Kubick, S. High-yield production of difficult-to-express proteins in a continuous exchange cell-free system based on CHO cell lysates. *Sci. Rep.* **2017**, *7*, 1–15. [CrossRef]

40. Hunt, J.P.; Yang, S.O.; Wilding, K.M.; Bundy, B.C. The growing impact of lyophilized cell-free protein expression systems. *Bioengineered* **2017**, *8*, 325–330. [CrossRef]

41. Madono, M.; Sawasaki, T.; Morishita, R.; Endo, Y. Wheat germ cell-free protein production system for post-genomic research. *N. Biotechnol.* **2011**, *28*, 211–217. [CrossRef]

42. Smith, M.T.; Berkheimer, S.D.; Werner, C.J.; Bundy, B.C. Lyophilized Escherichia coli-based cell-free systems for robust, high-density, long-term storage. *Biotechniques* **2014**, *56*, 186–193. [CrossRef]

43. Dopp, J.L.; Reuel, N.F. Process optimization for scalable E. coli extract preparation for cell-free protein synthesis. *Biochem. Eng. J.* **2018**, *138*, 21–28. [CrossRef]

44. Pardee, K.; Green, A.A.; Ferrante, T.; Cameron, D.E.; DaleyKeyser, A.; Yin, P.; Collins, J.J. Paper-based synthetic gene networks. *Cell* **2014**, *159*, 940–954. [CrossRef]

45. Pardee, K.; Green, A.A.; Takahashi, M.K.; Braff, D.; Lambert, G.; Lee, J.W.; Ferrante, T.; Ma, D.; Donghia, N.; Fan, M.; et al. Rapid, Low-Cost Detection of Zika Virus Using Programmable Biomolecular Components. *Cell* **2016**, *165*, 1255–1266. [CrossRef]

46. Stark, J.C.; Huang, A.; Nguyen, P.Q.; Dubner, R.S.; Hsu, K.J.; Ferrante, T.C.; Anderson, M.; Kanapskyte, A.; Mucha, Q.; Packett, J.S.; et al. BioBits™ Bright: A fluorescent synthetic biology education kit. *Sci. Adv.* **2018**, *4*, 1–11. [CrossRef]

47. Pardee, K.; Slomovic, S.; Nguyen, P.Q.; Lee, J.W.; Donghia, N.; Burrill, D.; Ferrante, T.; McSorley, F.R.; Furuta, Y.; Vernet, A.; et al. Portable, On-Demand Biomolecular Manufacturing. *Cell* **2016**, *167*, 248–259. [CrossRef]

48. Gale, B.; Jafek, A.; Lambert, C.; Goenner, B.; Moghimifam, H.; Nze, U.; Kamarapu, S. A Review of Current Methods in Microfluidic Device Fabrication and Future Commercialization Prospects. *Inventions* **2018**, *3*, 60. [CrossRef]

49. Damiati, S.; Mhanna, R.; Kodzius, R.; Ehmoser, E.K. Cell-free approaches in synthetic biology utilizing microfluidics. *Genes (Basel)* **2018**, *9*, 144. [CrossRef]

50. Gerber, D.; Maerkl, S.J.; Quake, S.R. An in vitro microfluidic approach to generating protein-interaction networks. *Nat. Methods* **2009**, *6*, 71–74. [CrossRef]

51. Jiao, Y.; Liu, Y.; Luo, D.; Huck, W.T.S.; Yang, D. Microfluidic-Assisted Fabrication of Clay Microgels for Cell-Free Protein Synthesis. *ACS Appl. Mater. Interfaces* **2018**, *10*, 29308–29313. [CrossRef]

52. Georgi, V.; Georgi, L.; Blechert, M.; Bergmeister, M.; Zwanzig, M.; Wüstenhagen, D.A.; Bier, F.F.; Jung, E.; Kubick, S. On-chip automation of cell-free protein synthesis: New opportunities due to a novel reaction mode. *Lab Chip* **2016**, *16*, 269–281. [CrossRef]

53. Huang, A.; Nguyen, P.Q.; Stark, J.C.; Takahashi, M.K.; Donghia, N.; Ferrante, T.; Dy, A.J.; Hsu, K.J.; Dubner, R.S.; Pardee, K.; et al. BioBits™ Explorer: A modular synthetic biology education kit. *Sci. Adv.* **2018**, *4*. [CrossRef]

54. Jaroentomeechai, T.; Stark, J.C.; Natarajan, A.; Glasscock, C.J.; Yates, L.E.; Hsu, K.J.; Mrksich, M.; Jewett, M.C.; Delisa, M.P. Single-pot glycoprotein biosynthesis using a cell-free transcription-translation system enriched with glycosylation machinery. *Nat. Commun.* **2018**, *9*.

55. Sanford, J.; Codina, J.; Birnbaumers, L. Gamma subunits of G Proteins, but not their alpha or beta subunits, are polyisoprenylated. *J. Biol. Chem.* **1991**, *266*, 9570–9579.

56. Dalley, J.A.; Bulleid, N.J. The Endoplasmic Reticulum (ER) Translocon can Differentiate between Hydrophobic Sequences Allowing Signals for Glycosylphosphatidylinositol Anchor Addition to be Fully Translocated into the ER Lumen. *J. Biol. Chem.* **2003**, *278*, 51749–51757. [CrossRef]

57. John, D.C.; Grant, M.E.; Bulleid, N.J. Cell-free synthesis and assembly of prolyl 4-hydroxylase: the role of the beta-subunit (PDI) in preventing misfolding and aggregation of the alpha-subunit. *EMBO J.* **1993**, *12*, 1587–1595. [CrossRef]

58. Gibbs, P.E.M.; Zouzias, D.C.; Freedberg, I.M. Differential post-translational modification of human type I keratins synthesized in a rabbit reticulocyte cell-free system. *Biochim. Biophys. Acta - Gene Struct. Expr.* **1985**, *824*, 247–255. [CrossRef]

59. Dougherty, W.G.; Dawn Parks, T. Post-translational processing of the tobacco etch virus 49-kDa small nuclear inclusion polyprotein: Identification of an internal cleavage site and delimitation of VPg and proteinase domains. *Virology* **1991**, *183*, 449–456. [CrossRef]

60. Starr, C.M.; Hanover, J.A. Glycosylation of nuclear pore protein p62. Reticulocyte lysate catalyzes O-linked N-acetylglucosamine addition in vitro. *J. Biol. Chem.* **1990**, *265*, 6868–6873.

61. Shields, D.; Blobel, G. Efficient Cleavage and Segregation of Nascent Presecretory Proteins in a Reticuloqyte Lysate Supplemented with Microsomal Membranes. *J. Biol. Chem.* **1978**, *253*, 3573–3576.

62. Ezure, T.; Suzuki, T.; Shikata, M.; Ito, M.; Ando, E.; Utsumi, T.; Nishimura, O.; Tsunasawa, S. Development of an insect cell-free system. *Curr Pharm Biotechnol* **2010**, *11*, 279–284. [CrossRef]

63. Tarui, H.; Murata, M.; Tani, I.; Imanishi, S.; Nishikawa, S.; Hara, T. Establishment and characterization of cell-free translation/glycosylation in insect cell (Spodoptera frugiperda 21) extract prepared with high pressure treatment. *Appl. Microbiol. Biotechnol.* **2001**, *55*, 446–453. [CrossRef]

64. Zemella, A.; Thoring, L.; Hoffmeister, C.; Kubick, S. Cell-Free Protein Synthesis: Pros and Cons of Prokaryotic and Eukaryotic Systems. *ChemBioChem* **2015**, *16*, 2420–2431. [CrossRef]

65. Suzuki, T.; Ito, M.; Ezure, T.; Shikata, M.; Ando, E.; Utsumi, T.; Tsunasawa, S.; Nishimura, O. Protein prenylation in an insect cell-free protein synthesis system and identification of products by mass spectrometry. *Proteomics* **2007**, *7*, 1942–1950. [CrossRef]

66. Suzuki, T.; Ezure, T.; Ando, E.; Nishimura, O.; Utsumi, T.; Tsunasawa, S. Preparation of ubiquitin-conjugated proteins using an insect cell-free protein synthesis system. *J. Biotechnol.* **2010**, *145*, 73–78. [CrossRef]

67. Katzen, F.; Kudlicki, W. Efficient generation of insect-based cell-free translation extracts active in glycosylation and signal sequence processing. *J. Biotechnol.* **2006**, *125*, 194–197. [CrossRef]

68. Zampatis, D.E.; Rutz, C.; Furkert, J.; Schmidt, A.; Wüstenhagen, D.; Kubick, S.; Tsopanoglou, N.E.; Schülein, R. The protease-activated receptor 1 possesses a functional and cleavable signal peptide which is necessary for receptor expression. *FEBS Lett.* **2012**, *586*, 2351–2359. [CrossRef]

69. von Groll, U.; Kubick, S.; Merk, H.; Stiege, W.; F, S. Advances in insect-based cell-free protein expression. In *Landes Bioscience*; Kudlicki, W., Katzen, F., Bennett, P., Eds.; Taylor & Francis: Austin, TX, USA, 2007; ISBN 1734998075.

70. Suzuki, T.; Ito, M.; Ezure, T.; Shikata, M.; Ando, E.; Utsumi, T.; Tsunasawa, S.; Nishimura, O. N-Terminal protein modifications in an insect cell-free protein synthesis system and their identification by mass spectrometry. *Proteomics* **2006**, *6*, 4486–4495. [CrossRef]

71. Ezure, T.; Suzuki, T.; Shikata, M.; Ito, M.; Ando, E.; Nishimura, O.; Tsunasawa, S. Expression of proteins containing disulfide bonds in an insect cell-free system and confirmation of their arrangements by MALDI-TOF MS. *Proteomics* **2007**, *7*, 4424–4434. [CrossRef]

72. Suzuki, T.; Moriya, K.; Nagatoshi, K.; Ota, Y.; Ezure, T.; Ando, E.; Tsunasawa, S.; Utsumi, T. Strategy for comprehensive identification of human N-myristoylated proteins using an insect cell-free protein synthesis system. *Proteomics* **2010**, *10*, 1780–1793. [CrossRef]

73. Sachse, R.; Wüstenhagen, D.; Šamalíková, M.; Gerrits, M.; Bier, F.F.; Kubick, S. Synthesis of membrane proteins in eukaryotic cell-free systems. *Eng. Life Sci.* **2013**, *13*, 39–48. [CrossRef]

74. Guarino, C.; DeLisa, M.P. A prokaryote-based cell-free translation system that efficiently synthesizes glycoproteins. *Glycobiology* **2012**, *22*, 596–601. [CrossRef]

75. Schoborg, J.A.; Hershewe, J.M.; Stark, J.C.; Kightlinger, W.; Kath, J.E.; Jaroentomeechai, T.; Natarajan, A.; DeLisa, M.P.; Jewett, M.C. A cell-free platform for rapid synthesis and testing of active oligosaccharyltransferases. *Biotechnol. Bioeng.* **2018**, *115*, 739–750. [CrossRef]

76. Rothblatt, J.A.; Meyer, D.I. Secretion in Yeast: Reconstitution of the Translocation and Glycosylation of a-Factor and Invertase in a Homologous Cell-Free System. *Cell* **1986**, *44*, 619–628. [CrossRef]

77. Shields, D.; Blobel, G. Cell-free synthesis of fish preproinsulin, and processing by heterologous mammalian microsomal membranes (mRNA from islets of Langerhans/wheat germ system/canine pancreatic microsomal membranes/amino-terminal sequences/ sequence homologies). *Cell Biol.* **1977**, *74*, 2059–2063.

78. Mikami, S.; Kobayashi, T.; Yokoyama, S.; Imataka, H. A hybridoma-based in vitro translation system that efficiently synthesizes glycoproteins. *J. Biotechnol.* **2006**, *127*, 65–78. [CrossRef]

79. Morita, E.H.; Sawasaki, T.; Tanaka, R.; Endo, Y.; Kohno, T. A wheat germ cell-free system is a novel way to screen protein folding and function. *Protein Sci.* **2003**, *12*, 1216–1221. [CrossRef]

80. Goshima, N.; Kawamura, Y.; Fukumoto, A.; Miura, A.; Honma, R.; Satoh, R.; Wakamatsu, A.; Yamamoto, J.I.; Kimura, K.; Nishikawa, T.; et al. Human protein factory for converting the transcriptome into an in vitro-expressed proteome. *Nat. Methods* **2008**, *5*, 1011–1017. [CrossRef]

81. Bryan, T.M.; Goodrich, K.J.; Cech, T.R. Telomerase RNA bound by protein motifs specific to telomerase reverse transcriptase. *Mol. Cell* **2000**, *6*, 493–499. [CrossRef]

82. Shao, J.; Prince, T.; Hartson, S.D.; Matts, R.L. Phosphorylation of Serine 13 Is Required for the Proper Function of the Hsp90 Co-chaperone, Cdc37. *J. Biol. Chem.* **2003**, *278*, 38117–38120. [CrossRef]

83. Keefe, A.D.; Szostak, J.W. Functional proteins from a random-sequence library. *Nature* **2001**, *410*, 715–718. [CrossRef]

84. Roberts, R.W.; Szostak, J.W. RNA-peptide fusions for the in vitro selection of peptides and proteins. *Proc. Natl. Acad. Sci. USA* **1997**, *94*, 12297–12302. [CrossRef]

85. Thoring, L.; Wustenhagen, D.A.; Borowiak, M.; Stech, M.; Sonnabend, A.; Kubick, S. Cell-free systems based on CHO cell lysates: Optimization strategies, synthesis of "difficult-to-express" proteins and future perspectives. *PLoS ONE* **2016**, *11*, e0163670. [CrossRef]

86. Bundy, B.C.; Swartz, J.R. Efficient disulfide bond formation in virus-like particles. *J. Biotechnol.* **2011**, *154*, 230–239. [CrossRef]

87. Bundy, B.C.; Franciszkowicz, M.J.; Swartz, J.R. Escherichia coli-based cell-free synthesis of virus-like particles. *Biotechnol. Bioeng.* **2008**, *100*, 28–37. [CrossRef]

88. Smith, M.T.; Varner, C.T.; Bush, D.B.; Bundy, B.C. The incorporation of the A2 protein to produce novel Qβ virus-like particles using cell-free protein synthesis. *Biotechnol. Prog.* **2012**, *28*, 549–555. [CrossRef]

89. Patel, K.G.; Swartz, J.R. Surface Functionalization of Virus-Like Particles by Direct Conjugation Using Azide−Alkyne Click Chemistry. *Bioconjug. Chem.* **2011**, *22*, 376–387. [CrossRef]

90. Rustad, M.; Eastlund, A.; Jardine, P.; Noireaux, V. Cell-free TXTL synthesis of infectious bacteriophage T4 in a single test tube reaction. *Synth. Biol.* **2018**, *3*. [CrossRef]

91. Franco, D.; Pathak, H.B.; Cameron, C.E.; Rombaut, B.; Wimmer, E.; Paul, A.V. Stimulation of Poliovirus Synthesis in a HeLa Cell-Free In Vitro Translation-RNA Replication System by Viral Protein 3CDpro. *J. Virol.* **2005**, *79*, 6358–6367. [CrossRef]

92. Spearman, P.; Ratner, L. Human immunodeficiency virus type 1 capsid formation in reticulocyte lysates. *J. Virol.* **1996**, *70*, 8187–8194.

93. Novelli, A.; Boulanger, P.A. Assembly of adenovirus type 2 fiber synthesized in cell-free translation system. *J. Biol. Chem.* **1991**, *266*, 9299–9303.

94. Klein, K.C.; Polyak, S.J.; Lingappa, J.R. Unique Features of Hepatitis C Virus Capsid Formation Revealed by De Novo Cell-Free Assembly. *J. Virol.* **2004**, *78*, 9257–9269. [CrossRef]

95. Wang, X.; Liu, J.; Zheng, Y.; Li, J.; Wang, H.; Zhou, Y.; Qi, M.; Yu, H.; Tang, W.; Zhao, W.M. An optimized yeast cell-free system: Sufficient for translation of human papillomavirus 58 L1 mRNA and assembly of virus-like particles. *J. Biosci. Bioeng.* **2008**, *106*, 8–15. [CrossRef]

96. Wang, X.; Liu, J.; Zhao, W.M.; Zhao, K.N. Translational comparison of HPV58 long and short L1 mRNAs in yeast (Saccharomyces cerevisiae) cell-free system. *J. Biosci. Bioeng.* **2010**, *110*, 58–65. [CrossRef]

97. Zawada, J.F.; Yin, G.; Steiner, A.R.; Yang, J.; Naresh, A.; Roy, S.M.; Gold, D.S.; Heinsohn, H.G.; Murray, C.J. Microscale to manufacturing scale-up of cell-free cytokine production-a new approach for shortening protein production development timelines. *Biotechnol. Bioeng.* **2011**, *108*, 1570–1578. [CrossRef]

98. Jackson, R.J.; Hunt, T. Preparation and use of nuclease-treated rabbit reticulocyte lysates for the translation of eukaryotic messenger RNA. *Methods Enzymol.* **1983**, *96*, 50–74.

99. Beebe, E.T.; Makino, S.I.; Nozawa, A.; Matsubara, Y.; Frederick, R.O.; Primm, J.G.; Goren, M.A.; Fox, B.G. Robotic large-scale application of wheat cell-free translation to structural studies including membrane proteins. *N. Biotechnol.* **2011**, *28*, 239–249. [CrossRef]

100. Schoborg, J.A.; Jewett, M.C. Cell-free protein synthesis: An emerging technology for understanding, harnessing, and expanding the capabilities of biological systems. In *Synthetic Biology: Parts, Devices and Applications*; John Wiley & Sons: Hoboken, NJ, USA, 2018; pp. 309–330. ISBN 9783527688104.

101. Oza, J.P.; Aerni, H.R.; Pirman, N.L.; Barber, K.W.; ter Haar, C.M.; Rogulina, S.; Amrofell, M.B.; Isaacs, F.J.; Rinehart, J.; Jewett, M.C. Robust production of recombinant phosphoproteins using cell-free protein synthesis. *Nat. Commun.* **2015**, *6*, 8168. [CrossRef]

102. Voloshin, A.M.; Swartz, J.R. Efficient and scalable method for scaling up cell free protein synthesis in batch mode. *Biotechnol. Bioeng.* **2005**, *91*, 516–521. [CrossRef]

103. Martin, R.W.; Majewska, N.I.; Chen, C.X.; Albanetti, T.E.; Jimenez, R.B.C.; Schmelzer, A.E.; Jewett, M.C.; Roy, V. Development of a CHO-Based Cell-Free Platform for Synthesis of Active Monoclonal Antibodies. *ACS Synth. Biol.* **2017**, *6*, 1370–1379. [CrossRef]

104. Jewett, M.C.; Swartz, J.R. Mimicking the Escherichia coli Cytoplasmic Environment Activates Long-Lived and Efficient Cell-Free Protein Synthesis. *Biotechnol. Bioeng.* **2004**, *86*, 19–26. [CrossRef]

105. Rigaud, J.L. Membrane proteins: Functional and structural studies using reconstituted proteoliposomes and 2-D crystals. *Brazilian J. Med. Biol. Res.* **2002**, *35*, 753–766. [CrossRef]

106. Sachse, R.; Dondapati, S.K.; Fenz, S.F.; Schmidt, T.; Kubick, S. Membrane protein synthesis in cell-free systems: From bio-mimetic systems to bio-membranes. *FEBS Lett.* **2014**, *588*, 2774–2781. [CrossRef]

107. Yadavalli, R.; Sam-Yellowe, T. HeLa Based Cell Free Expression Systems for Expression of Plasmodium Rhoptry Proteins. *J. Vis. Exp.* **2015**, e52772. [CrossRef]

108. Dondapati, S.K.; Kreir, M.; Quast, R.B.; Wüstenhagen, D.A.; Brüggemann, A.; Fertig, N.; Kubick, S. Membrane assembly of the functional KcsA potassium channel in a vesicle-based eukaryotic cell-free translation system. *Biosens. Bioelectron.* **2014**, *59*, 174–183. [CrossRef]

109. Wilson, C.M.; Farmery, M.R.; Bulleid, N.J. Pivotal role of calnexin and mannose trimming in regulating the endoplasmic reticulum-associated degradation of major histocompatibility complex class I heavy chain. *J. Biol. Chem.* **2000**, *275*, 21224–21232. [CrossRef]

110. Goren, M.A.; Fox, B.G. Wheat germ cell-free translation, purification, and assembly of a functional human stearoyl-CoA desaturase complex. *Protein Expr. Purif.* **2008**, *62*, 171–178. [CrossRef]

111. Nozawa, A.; Nanamiya, H.; Miyata, T.; Linka, N.; Endo, Y.; Weber, A.P.M.; Tozawa, Y. A cell-free translation and proteoliposome reconstitution system for functional analysis of plant solute transporters. *Plant Cell Physiol.* **2007**, *48*, 1815–1820. [CrossRef]

112. Junge, F.; Haberstock, S.; Roos, C.; Stefer, S.; Proverbio, D.; Dötsch, V.; Bernhard, F. Advances in cell-free protein synthesis for the functional and structural analysis of membrane proteins. *N. Biotechnol.* **2011**, *28*, 262–271. [CrossRef]

113. Arduengo, M.; Schenborn, E.; Hurst, R. *The Role of Cell-Free Rabbit Reticulocyte Expression Systems in Functional Proteomics. Cell-free Expression*; Landes Bioscience: Austin, TX, USA, 2007; pp. 1–18.

114. Stavnezer, J.; Huang, R.C.C. Synthesis of a Mouse Immunoglobulin Light Chain in a Rabbit Reticulocyte Cell-free System. *Nat. New Biol.* **1971**, *230*, 172–176. [CrossRef]

115. Nicholls, P.J.; Johnson, V.G.; Andrew, S.M.; Hoogenboom, H.R.; Raus, J.C.; Youle, R.J. Characterization of single-chain antibody (sFv)-toxin fusion proteins produced in vitro in rabbit reticulocyte lysate. *J. Biol. Chem.* **1993**, *268*, 5302–5308.

116. Gusdon, B.Y.J.P.; Stavitsky, A.B.; Armentrout, S.A. Synthesis of gamma G antibody and immunoglobulin on polyribosomes in a cell-free system. *Proc. Natl. Acad. Sci.* **1967**, *58*, 1189–1196. [CrossRef]

117. Ryabova, L.A.; Desplancq, D.; Spirin, A.S.; Plückthun, A. Functional antibody production using cell-free translation: Effects of protein disulfide isomerase and chaperones. *Nat. Biotechnol.* **1997**, *15*, 79–84. [CrossRef]

118. Merk, H.; Stiege, W.; Tsumoto, K.; Kumagai, I.; Erdmann, V.A. Cell-free expression of two single-chain monoclonal antibodies against lysozyme: effect of domain arrangement on the expression. *J. Biochem.* **1999**, *125*, 328–333. [CrossRef]

119. Zimmerman, E.S.; Heibeck, T.H.; Gill, A.; Li, X.; Murray, C.J.; Madlansacay, M.R.; Tran, C.; Uter, N.T.; Yin, G.; Rivers, P.J.; et al. Production of site-specific antibody-drug conjugates using optimized non-natural amino acids in a cell-free expression system. *Bioconjug. Chem.* **2014**, *25*, 351–361. [CrossRef]

120. Galeffi, P.; Lombardi, A.; Pietraforte, I.; Novelli, F.; Di Donato, M.; Sperandei, M.; Tornambé, A.; Fraioli, R.; Martayan, A.; Natali, P.G.; et al. Functional expression of a single-chain antibody to ErbB-2 in plants and cell-free systems. *J. Transl. Med.* **2006**, *4*, 39. [CrossRef]

121. Jiang, X.; Ookubo, Y.; Fujii, I.; Nakano, H.; Yamane, T. Expression of Fab fragment of catalytic antibody 6D9 in an Escherichia coli in vitro coupled transcription/translation system. *FEBS Lett.* **2002**, *514*, 290–294. [CrossRef]

122. Groff, D.; Armstrong, S.; Rivers, P.J.; Zhang, J.; Yang, J.; Green, E.; Rozzelle, J.; Liang, S.; Kittle, J.D.; Steiner, A.R.; et al. Engineering toward a bacterial "endoplasmic reticulum" for the rapid expression of immunoglobulin proteins. *MAbs* **2014**, *6*, 671–678. [CrossRef]

123. Kawasaki, T.; Gouda, M.D.; Sawasaki, T.; Takai, K.; Endo, Y. Efficient synthesis of a disulfide-containing protein through a batch cell-free system from wheat germ. *Eur. J. Biochem.* **2003**, *270*, 4780–4786. [CrossRef]

124. Stech, M.; Merk, H.; Schenk, J.A.; Stöcklein, W.F.M.; Wüstenhagen, D.A.; Micheel, B.; Duschl, C.; Bier, F.F.; Kubick, S. Production of functional antibody fragments in a vesicle-based eukaryotic cell-free translation system. *J. Biotechnol.* **2012**, *164*, 220–231. [CrossRef]

125. Stech, M.; Hust, M.; Schulze, C.; Dübel, S.; Kubick, S. Cell-free eukaryotic systems for the production, engineering, and modification of scFv antibody fragments. *Eng. Life Sci.* **2014**, *14*, 387–398. [CrossRef]

126. Goering, A.W.; Li, J.; McClure, R.A.; Thomson, R.J.; Jewett, M.C.; Kelleher, N.L. In vitro reconstruction of nonribosomal peptide biosynthesis directly from DNA using cell-free protein synthesis. *ACS Synth. Biol.* **2017**, *6*, 39–44. [CrossRef]

127. Mikami, S.; Kobayashi, T.; Masutani, M.; Yokoyama, S.; Imataka, H. A human cell-derived in vitro coupled transcription/translation system optimized for production of recombinant proteins. *Protein Expr. Purif.* **2008**, *62*, 190–198. [CrossRef] [PubMed]

128. Matsumura, Y.; Rooney, L.; Skach, W.R. In vitro methods for CFTR biogenesis. *Methods Mol. Biol.* **2011**, *741*, 233–253.

129. Brödel, A.K.; Raymond, J.A.; Duman, J.G.; Bier, F.F.; Kubick, S. Functional evaluation of candidate ice structuring proteins using cell-free expression systems. *J. Biotechnol.* **2013**, *163*, 301–310. [CrossRef] [PubMed]

130. Li, J.; Lawton, T.J.; Kostecki, J.S.; Nisthal, A.; Fang, J.; Mayo, S.L.; Rosenzweig, A.C.; Jewett, M.C. Cell-free protein synthesis enables high yielding synthesis of an active multicopper oxidase. *Biotechnol. J.* **2016**, *11*, 212–218. [CrossRef] [PubMed]

131. Boyer, M.E.; Stapleton, J.A.; Kuchenreuther, J.M.; Wang, C.; Swartz, J.R. Cell-free synthesis and maturation of [FeFe] hydrogenases. *Biotechnol. Bioeng.* **2008**, *99*, 59–67. [CrossRef] [PubMed]

132. Kuchenreuther, J.M.; Shiigi, S.A.; Swartz, J.R. Cell-Free Synthesis of the H-Cluster: A Model for the In Vitro Assembly of Metalloprotein Metal Centers. In *Methods in molecular biology (Clifton, N.J.)*; Humana Press: Totowa, NJ, USA, 2014; Volume 1122, pp. 49–72.

133. Ahn, J.-H.; Hwang, M.-Y.; Oh, I.-S.; Park, K.-M.; Hahn, G.-H.; Choi, C.-Y.; Kim, D.-M. Preparation method forEscherichia coli S30 extracts completely dependent upon tRNA addition to catalyze cell-free protein synthesis. *Biotechnol. Bioprocess Eng.* **2006**, *11*, 420–424. [CrossRef]

134. Yokogawa, T.; Kitamura, Y.; Nakamura, D.; Ohno, S.; Nishikawa, K. Optimization of the hybridization-based method for purification of thermostable tRNAs in the presence of tetraalkylammonium salts. *Nucleic Acids Res.* **2010**, *38*, e89. [CrossRef] [PubMed]

135. Panthu, B.; Ohlmann, T.; Perrier, J.; Schlattner, U.; Jalinot, P.; Elena-Herrmann, B.; Rautureau, G.J.P. Cell-Free Protein Synthesis Enhancement from Real-Time NMR Metabolite Kinetics: Redirecting Energy Fluxes in Hybrid RRL Systems. *ACS Synth. Biol.* **2018**, *7*, 218–226. [CrossRef] [PubMed]

136. Jewett, M.C.; Fritz, B.R.; Timmerman, L.E.; Church, G.M. In vitro integration of ribosomal RNA synthesis, ribosome assembly, and translation. *Mol. Syst. Biol.* **2013**, *9*, 1–8. [CrossRef] [PubMed]

137. Timm, A.C.; Shankles, P.G.; Foster, C.M.; Doktycz, M.J.; Retterer, S.T. Toward Microfluidic Reactors for Cell-Free Protein Synthesis at the Point-of-Care. *Small* **2016**, *12*, 810–817. [CrossRef] [PubMed]

138. Brophy, J.A.N.; Voigt, C.A. Principles of genetic circuit design. *Nat. Methods* **2014**, *11*, 508–520. [CrossRef] [PubMed]

139. Noireaux, V.; Bar-Ziv, R.; Libchaber, A. Principles of cell-free genetic circuit assembly. *Proc. Natl. Acad. Sci.* **2003**, *100*, 12672–12677. [CrossRef] [PubMed]

140. Karig, D.K.; Jung, S.-Y.; Srijanto, B.; Collier, C.P.; Simpson, M.L. Probing Cell-Free Gene Expression Noise in Femtoliter Volumes. *ACS Synth. Biol.* **2013**, *2*, 497–505. [CrossRef] [PubMed]

141. Siegal-Gaskins, D.; Noireaux, V.; Murray, R.M. Biomolecular resource utilization in elementary cell-free gene circuits. In Proceedings of the 2013 American Control Conference, Washington, DC, USA, 17–19 June 2013.

142. Karzbrun, E.; Shin, J.; Bar-Ziv, R.H.; Noireaux, V. Coarse-grained dynamics of protein synthesis in a cell-free system. *Phys. Rev. Lett.* **2011**, *106*. [CrossRef] [PubMed]

143. Shin, J.; Noireaux, V. An E. coli cell-free expression toolbox: Application to synthetic gene circuits and artificial cells. *ACS Synth. Biol.* **2012**, *1*, 29–41. [CrossRef] [PubMed]

144. Maerkl, S.J.; Murray, R.M.; Sun, Z.Z.; Niederholtmeyer, H.; Hori, Y.; Yeung, E.; Verpoorte, A. Rapid cell-free forward engineering of novel genetic ring oscillators. *Elife* **2015**, *4*, 1–18.

145. Noireaux, V.; Singhal, V.; Sun, Z.Z.; Murray, R.M.; Spring, K.J.; Chappell, J.; Hayes, C.A.; Lucks, J.B.; Fall, C.P.; Al-Khabouri, S.; et al. Rapidly Characterizing the Fast Dynamics of RNA Genetic Circuitry with Cell-Free Transcription–Translation (TX-TL) Systems. *ACS Synth. Biol.* **2014**, *4*, 503–515.

146. Halleran, A.D.; Murray, R.M. Cell-Free and in Vivo Characterization of Lux, Las, and Rpa Quorum Activation Systems in E. coli. *ACS Synth. Biol.* **2018**, *7*, 752–755. [CrossRef] [PubMed]

147. Sen, S.; Apurva, D.; Satija, R.; Siegal, D.; Murray, R.M. Design of a Toolbox of RNA Thermometers. *ACS Synth. Biol.* **2017**, *6*, 1461–1470. [CrossRef]

148. Dudley, Q.M.; Anderson, K.C.; Jewett, M.C. Cell-Free Mixing of Escherichia coli Crude Extracts to Prototype and Rationally Engineer High-Titer Mevalonate Synthesis. *ACS Synth. Biol.* **2016**, *5*, 1578–1588. [CrossRef] [PubMed]

149. Jiang, L.; Zhao, J.; Lian, J.; Xu, Z. Cell-free protein synthesis enabled rapid prototyping for metabolic engineering and synthetic biology. *Synth. Syst. Biotechnol.* **2018**, *3*, 90–96. [CrossRef] [PubMed]

150. Khattak, W.A.; Ul-Islam, M.; Ullah, M.W.; Yu, B.; Khan, S.; Park, J.K. Yeast cell-free enzyme system for bio-ethanol production at elevated temperatures. *Process Biochem.* **2014**, *49*, 357–364. [CrossRef]

151. Kay, J.E.; Jewett, M.C. Lysate of engineered Escherichia coli supports high-level conversion of glucose to 2,3-butanediol. *Metab. Eng.* **2015**, *32*, 133–142. [CrossRef] [PubMed]

152. Quast, R.B.; Mrusek, D.; Hoffmeister, C.; Sonnabend, A.; Kubick, S. Cotranslational incorporation of non-standard amino acids using cell-free protein synthesis. *FEBS Lett.* **2015**, *589*, 1703–1712. [CrossRef] [PubMed]

153. Albayrak, C.; Swartz, J.R. Cell-free co-production of an orthogonal transfer RNA activates efficient site-specific non-natural amino acid incorporation. *Nucleic Acids Res.* **2013**, *41*, 5949–5963. [CrossRef] [PubMed]

154. Martin, R.W.; Des Soye, B.J.; Kwon, Y.-C.; Kay, J.; Davis, R.G.; Thomas, P.M.; Majewska, N.I.; Chen, C.X.; Marcum, R.D.; Weiss, M.G.; et al. Cell-free protein synthesis from genomically recoded bacteria enables multisite incorporation of noncanonical amino acids. *Nat. Commun.* **2018**, *9*, 1203. [CrossRef] [PubMed]

155. Ozer, E.; Chemla, Y.; Schlesinger, O.; Aviram, H.Y.; Riven, I.; Haran, G.; Alfonta, L. In vitro suppression of two different stop codons. *Biotechnol. Bioeng.* **2017**, *114*, 1065–1073. [CrossRef] [PubMed]

156. Benítez-Mateos, A.I.; Llarena, I.; Sánchez-Iglesias, A.; López-Gallego, F. Expanding One-Pot Cell-Free Protein Synthesis and Immobilization for On-Demand Manufacturing of Biomaterials. *ACS Synth. Biol.* **2018**, *7*, 875–884. [CrossRef] [PubMed]

157. Cui, Z.; Wu, Y.; Mureev, S.; Alexandrov, K. Oligonucleotide-mediated tRNA sequestration enables one-pot sense codon reassignment in vitro. *Nucleic Acids Res.* **2018**, *46*, 6387–6400. [CrossRef] [PubMed]

158. Gao, W.; Bu, N.; Lu, Y.; Gao, W.; Bu, N.; Lu, Y. Efficient Incorporation of Unnatural Amino Acids into Proteins with a Robust Cell-Free System. *Methods Protoc.* **2019**, *2*, 16. [CrossRef]

159. Quast, R.B.; Kortt, O.; Henkel, J.; Dondapati, S.K.; Wüstenhagen, D.A.; Stech, M.; Kubick, S. Automated production of functional membrane proteins using eukaryotic cell-free translation systems. *J. Biotechnol.* **2015**, *203*, 45–53. [CrossRef] [PubMed]

160. Stech, M.; Brödel, A.K.; Quast, R.B.; Sachse, R.; Kubick, S. Cell-Free Systems: Functional Modules for Synthetic and Chemical Biology. In *Advances in Biochemical Engineering/Biotechnology*; Springer: Berlin/Heidelberg, Germany, 2013; Volume 137, pp. 67–102.

161. Wu, C.; Dasgupta, A.; Shen, L.; Bell-Pedersen, D.; Sachs, M.S. The cell free protein synthesis system from the model filamentous fungus Neurospora crassa. *Methods* **2018**, *137*, 11–19. [CrossRef]

162. Szczesna-Skorupa, E.; Filipowicz, W.; Paszewski, A. The cell-free protein synthesis system from the "slime" mutant of Neurospora crassa. Preparation and characterisation of importance of 7-methylguanosine for translation of viral and cellular mRNAs. *Eur. J. Biochem.* **1981**, *121*, 163–168. [CrossRef] [PubMed]

163. Curle, C.A.; Kapoor, M. A Neurospora crassa heat-shocked cell lysate translates homologous and heterologous messenger RNA efficiently, without preference for heat shock messages. *Curr. Genet.* **1988**, *13*, 401–409. [CrossRef] [PubMed]

164. Sachs, M.S.; Wang, Z.; Gaba, A.; Fang, P.; Belk, J.; Ganesan, R.; Amrani, N.; Jacobson, A. Toeprint analysis of the positioning of translation apparatus components at initiation and termination codons of fungal mRNAs. *Methods* **2002**, *26*, 105–114. [CrossRef]

165. Wu, C.; Wei, J.; Lin, P.-J.; Tu, L.; Deutsch, C.; Johnson, A.E.; Sachs, M.S. Arginine Changes the Conformation of the Arginine Attenuator Peptide Relative to the Ribosome Tunnel. *J. Mol. Biol.* **2012**, *416*, 518–533. [CrossRef]

166. Wei, J.; Wu, C.; Sachs, M.S. The Arginine Attenuator Peptide Interferes with the Ribosome Peptidyl Transferase Center. *Mol. Cell. Biol.* **2012**, *32*, 2396–2406. [CrossRef]

167. Yu, C.-H.; Dang, Y.; Zhou, Z.; Wu, C.; Zhao, F.; Sachs, M.S.; Liu, Y. Codon Usage Influences the Local Rate of Translation Elongation to Regulate Co-translational Protein Folding. *Mol. Cell* **2015**, *59*, 744–754. [CrossRef]

168. Li, J.; Wang, H.; Kwon, Y.C.; Jewett, M.C. Establishing a high yielding streptomyces-based cell-free protein synthesis system. *Biotechnol. Bioeng.* **2017**, *114*, 1343–1353. [CrossRef] [PubMed]

169. Thompson, J.; Rae, S.; Cundliffe, E. Coupled transcription–translation in extracts of Streptomyces lividans. *Mol. Gen. Genet.* **1984**, *195*, 39–43. [CrossRef]

170. Wiegand, D.J.; Lee, H.H.; Ostrov, N.; Church, G.M. Establishing a Cell-Free Vibrio natriegens Expression System. *ACS Synth. Biol.* **2018**, *7*, 2475–2479. [CrossRef]

171. Failmezger, J.; Scholz, S.; Blombach, B.; Siemann-Herzberg, M. Cell-free protein synthesis from fast-growing Vibrio natriegens. *Front. Microbiol.* **2018**, *9*. [CrossRef]

172. Tominaga, A.; Kobayashi, Y. Kasugamycin-resistant mutants of Bacillus subtilis. *J. Bacteriol.* **1978**, *135*, 1149–1150.

173. Stallcup, M.R.; Sharrock, W.J.; Rabinowitz, J.C. Specificity of bacterial ribosomes and messenger ribonucleic acids in protein synthesis reactions in vitro. *J. Biol. Chem.* **1976**, *251*, 2499–2510.

174. Villafane, R.; Bechhofer, D.H.; Narayanan, C.S.; Dubnau, D. Replication control genes of plasmid pE194. *J. Bacteriol.* **1987**, *169*, 4822–4829. [CrossRef]

175. Spencer, D.; Wildman, S.G. The Incorporation of Amino Acids into Protein by Cell-free Extracts from Tobacco Leaves. *Biochemistry* **1964**, *3*, 954–959. [CrossRef]

176. Boardman, N.K.; Francki, R.I.B.; Wildman, S.G. Protein synthesis by cell-free extracts of tobacco leaves: III. Comparison of the physical properties and protein synthesizing activities of 70 s chloroplast and 80 s cytoplasmic ribosomes. *J. Mol. Biol.* **1966**, *17*, 470–487. [CrossRef]

177. Phelps, R.H.; Sequeira, L. Synthesis of Indoleacetic Acid via Tryptamine by a Cell-free System from Tobacco Terminal Buds. *Plant Physiol.* **1967**, *42*, 1161–1163. [CrossRef]

178. Guo, Z.H.; Severson, R.F.; Wagner, G.J. Biosynthesis of the Diterpene cis-Abienol in Cell-Free Extracts of Tobacco Trichomes. *Arch. Biochem. Biophys.* **1994**, *308*, 103–108. [CrossRef]

179. Cooke, R.; Penon, P. In vitro transcription from cauliflower mosaic virus promoters by a cell-free extract from tobacco cells. *Plant Mol. Biol.* **1990**, *14*, 391–405. [CrossRef]

180. Chen, C.; Melitz, D.K. Cytokinin biosynthesis in a cell-free system from cytokinin-autotrophic tobacco tissue cultures. *FEBS Lett.* **1979**, *107*, 15–20. [CrossRef]

181. Hao, D.; Yeoman, M.M. Nicotine N-demethylase in cell-free preparations from tobacco cell cultures. *Phytochemistry* **1996**, *42*, 325–329. [CrossRef]

182. Komoda, K.; Naito, S.; Ishikawa, M. Replication of plant RNA virus genomes in a cell-free extract of evacuolated plant protoplasts. *Proc. Natl. Acad. Sci.* **2004**, *101*, 1863–1867. [CrossRef]

183. Gursinsky, T.; Schulz, B.; Behrens, S.E. Replication of Tomato bushy stunt virus RNA in a plant in vitro system. *Virology* **2009**, *390*, 250–260. [CrossRef]

184. Murota, K.; Hagiwara-Komoda, Y.; Komoda, K.; Onouchi, H.; Ishikawa, M.; Naito, S. Arabidopsis cell-free extract, ACE, a new in vitro translation system derived from arabidopsis callus cultures. *Plant Cell Physiol.* **2011**, *52*, 1443–1453. [CrossRef]

185. Moore, S.J.; MacDonald, J.T.; Wienecke, S.; Ishwarbhai, A.; Tsipa, A.; Aw, R.; Kylilis, N.; Bell, D.J.; McClymont, D.W.; Jensen, K.; et al. Rapid acquisition and model-based analysis of cell-free transcription–translation reactions from nonmodel bacteria. *Proc. Natl. Acad. Sci.* **2018**, 201715806. [CrossRef]

186. Ruggero, D.; Creti, R.; Londei, P. In vitro translation of archaeal natural mRNAs at high temperature. *FEMS Microbiol. Lett.* **1993**, *107*, 89–94. [CrossRef]

187. Elhardt, D.; Böck, A. An in vitro polypeptide synthesizing system from methanogenic bacteria: Sensitivity to antibiotics. *MGG Mol. Gen. Genet.* **1982**, *188*, 128–134. [CrossRef]

188. Uzawa, T.; Yamagishi, A.; Oshima, T. Polypeptide synthesis directed by DNA as a messenger in cell-free polypeptide synthesis by extreme thermophiles, Thermus thermophilus HB27 and sulfolobus tokodaii strain 7. *J. Biochem.* **2002**, *131*, 849–853. [CrossRef]

189. Uzawa, T.; Hamasaki, N.; Oshima, T. Effects of Novel Polyamines on Cell-Free Polypeptide Catalyzed by Thermus thermophilus HB8 Extract. *J. Biochem.* **1993**, *486*, 478–486. [CrossRef]

190. Kovtun, O.; Mureev, S.; Jung, W.R.; Kubala, M.H.; Johnston, W.; Alexandrov, K. Leishmania cell-free protein expression system. *Methods* **2011**, *55*, 58–64. [CrossRef]

191. Bhide, M.; Natarajan, S.; Hresko, S.; Aguilar, C.; Bencurova, E. Rapid in vitro protein synthesis pipeline: A promising tool for cost-effective protein array design. *Mol. Biosyst.* **2014**, *10*, 1236–1245. [CrossRef]

192. Perez, J.G.; Stark, J.C.; Jewett, M.C. Cell-Free Synthetic Biology: Engineering Beyond the Cell. *Cold Spring Harb. Perspect. Biol.* **2016**, *8*, 1–26. [CrossRef]

193. Ruehrer, S.; Michel, H. Exploiting Leishmania tarentolae cell-free extracts for the synthesis of human solute carriers. *Mol. Membr. Biol.* **2013**, *30*, 288–302. [CrossRef]

194. Madin, K.; Sawasaki, T.; Ogasawara, T.; Endo, Y. A highly efficient and robust cell-free protein synthesis system prepared from wheat embryos: plants apparently contain a suicide system directed at ribosomes. *Proc. Natl. Acad. Sci. USA* **2000**, *97*, 559–664. [CrossRef]

195. Hodgman, C.E.; Jewett, M.C. Optimized extract preparation methods and reaction conditions for improved yeast cell-free protein synthesis. *Biotechnol. Bioeng.* **2013**, *110*, 2643–2654. [CrossRef]

196. Kubick, S.; Schacherl, J.; Fleisher-Notter, H.; Royall, E.; Roberts, L.O.; Steige, W. In Vitro Translation in an Insect-Based Cell-Free System. In *Cell-free Protein Expression*; Springer: Berlin/Heidelberg, Germany, 2003; ISBN 978-3-642-63939-5.

197. Kwon, Y.-C.; Jewett, M.C. High-throughput preparation methods of crude extract for robust cell-free protein synthesis. *Sci. Rep.* **2015**, *5*, 8663. [CrossRef]

198. Krinsky, N.; Kaduri, M.; Shainsky-Roitman, J.; Goldfeder, M.; Ivanir, E.; Benhar, I.; Shoham, Y.; Schroeder, A. A Simple and Rapid Method for Preparing a Cell-Free Bacterial Lysate for Protein Synthesis. *PLoS ONE* **2016**, *11*, e0165137. [CrossRef]

199. Shin, J.; Noireaux, V. Efficient cell-free expression with the endogenous E. Coli RNA polymerase and sigma factor 70. *J. Biol. Eng.* **2010**, *4*, 8. [CrossRef]

200. Nishimura, N.; Kitaoka, Y.; Mimura, A.; Takahara, Y. Continuous protein synthesis system with Escherichia coli S30 extract containing endogenous T7 RNA polymerase. *Biotechnol. Lett.* **1993**, *15*, 785–790. [CrossRef]

201. Lee, K.-H.; Kim, D.-M. Recent advances in development of cell-free protein synthesis systems for fast and efficient production of recombinant proteins. *FEMS Microbiol. Lett.* **2018**, *365*, 1–7. [CrossRef] [PubMed]

202. Condò, I.; Ciammaruconi, A.; Benelli, D.; Ruggero, D.; Londei, P. Cis-acting signals controlling translational initiation in the thermophilic archaeon Sulfolobus solfataricus. *Mol. Microbiol.* **1999**, *34*, 377–384. [CrossRef] [PubMed]

203. Brödel, A.K.; Sonnabend, A.; Roberts, L.O.; Stech, M.; Wüstenhagen, D.A.; Kubick, S. IRES-mediated translation of membrane proteins and glycoproteins in eukaryotic cell-free systems. *PLoS ONE* **2013**, *8*, e82234.

204. Anastasina, M.; Terenin, I.; Butcher, S.J.; Kainov, D.E. A technique to increase protein yield in a rabbit reticulocyte lysate translation system. *Biotechniques* **2014**, *56*, 36–39. [CrossRef] [PubMed]

205. Hodgman, C.E.; Jewett, M.C. Characterizing IGR IRES-mediated translation initiation for use in yeast cell-free protein synthesis. *N. Biotechnol.* **2014**, *31*, 499–505. [CrossRef] [PubMed]

206. Dopp, B.J.L.; Tamiev, D.D.; Reuel, N.F. Cell-free supplement mixtures: Elucidating the history and biochemical utility of additives used to support in vitro protein synthesis in E. coli extract. *Biotechnol. Adv.* **2019**, *37*, 246–258. [CrossRef] [PubMed]

207. Li, J.; Gu, L.; Aach, J.; Church, G.M. Improved cell-free RNA and protein synthesis system. *PLoS ONE* **2014**, *9*, e106232. [CrossRef] [PubMed]

208. Zhang, Y.; Huang, Q.; Deng, Z.; Xu, Y.; Liu, T. Enhancing the efficiency of cell-free protein synthesis system by systematic titration of transcription and translation components. *Biochem. Eng. J.* **2018**, *138*, 47–53. [CrossRef]

![methods and protocols logo] *methods and protocols*

MDPI

Review

Cell-Free Synthetic Biology Platform for Engineering Synthetic Biological Circuits and Systems

Dohyun Jeong [1,†], Melissa Klocke [2,†], Siddharth Agarwal [2], Jeongwon Kim [1], Seungdo Choi [1], Elisa Franco [3,*] and Jongmin Kim [1,*]

1   Division of Integrative Biosciences and Biotechnology, Pohang University of Science and Technology, 77 Cheongam-ro, Pohang, Gyeongbuk 37673, Korea; gyu9506@postech.ac.kr (D.J.); jeongwon96@postech.ac.kr (J.K.); choisd@postech.ac.kr (S.C.)
2   Department of Mechanical Engineering, University of California at Riverside, 900 University Ave, Riverside, CA 92521, USA; klocke@ucr.edu (M.K.); sagar002@ucr.edu (S.A.)
3   Department of Mechanical and Aerospace Engineering, University of California at Los Angeles, 420 Westwood Plaza, Los Angeles, CA 90095, USA
*   Correspondence: efranco@seas.ucla.edu (E.F.); jongmin.kim@postech.ac.kr (J.K.); Tel.: +1-310-206-4830 (E.F.); +82-54-279-2322 (J.K.)
†   Both authors contributed equally to this work.

Received: 4 March 2019; Accepted: 8 May 2019; Published: 14 May 2019

Abstract: Synthetic biology integrates diverse engineering disciplines to create novel biological systems for biomedical and technological applications. The substantial growth of the synthetic biology field in the past decade is poised to transform biotechnology and medicine. To streamline design processes and facilitate debugging of complex synthetic circuits, cell-free synthetic biology approaches has reached broad research communities both in academia and industry. By recapitulating gene expression systems in vitro, cell-free expression systems offer flexibility to explore beyond the confines of living cells and allow networking of synthetic and natural systems. Here, we review the capabilities of the current cell-free platforms, focusing on nucleic acid-based molecular programs and circuit construction. We survey the recent developments including cell-free transcription–translation platforms, DNA nanostructures and circuits, and novel classes of riboregulators. The links to mathematical models and the prospects of cell-free synthetic biology platforms will also be discussed.

Keywords: synthetic biology; cell-free transcription-translation; rapid prototyping; artificial cell; riboregulator; DNA origami; mathematical model

---

1. Introduction

Synthetic biology focuses on engineering biological circuits to manipulate biological systems and technological applications. Formative works in synthetic biology demonstrated the creation of simple regulatory circuits in *Escherichia coli* [1,2]. The dynamics of these synthetic circuits were reasonably captured through mathematical modeling, driving further developments of forward-engineering approaches [3]. As the scope of synthetic biological circuits increases dramatically, comprehensive design, analysis, and predictive modeling in cellular contexts becomes challenging despite progress in computer-aided designs [4,5]. Cell-free synthetic biology provides a paradigm to test components and circuits in a well-controlled environment that is similar to physiological conditions [6]. Cell-free approaches could expedite development and exploration of synthetic system designs beyond the confines of living organisms. In turn, novel, sustainable, and cost-effective technologies based on cell-free synthetic biology could help meet broader, worldwide challenges in the future.

In this article, we review the current scope of cell-free synthetic biology, focusing on synthetic circuits and systems using nucleic acid-based programs. We limit ourselves to the design and applications of these synthetic molecular circuits. Readers are referred to other excellent reviews for recent developments in other areas of cell-free synthetic biology such as cell-free metabolic engineering [7,8]. We first survey cell-free transcription–translation platforms that are gaining popularity as a testbed for rapid prototyping of synthetic circuit elements and circuitry. We then review in vitro model dynamical systems and recent progress in de novo-designed RNA regulatory toolkits for synthetic biology. Next, we discuss synthetic cell approaches through compartmentalization and the prospect of nucleic acid-based nanostructures and circuits to function in cell-like environments. Finally, we discuss modeling approaches and developments as well as their links with the future of synthetic biological circuits.

2. Cell-Free Transcription–Translation Platform for Synthetic Biology

Synthetic biology approaches for achieving novel and complex functionality in cellular systems have shown significant progress. Using cells as chassis to engineer circuits, however, presents challenges for rapid design–build–test cycles despite ongoing development of applicable tools. The cell-free transcription–translation system presents an attractive alternative to construct, characterize, and interrogate synthetic biological circuits (Figure 1). Although a number of cell-free expression systems have been developed, including rabbit reticulocytes, wheat germ, and insect cells, the prokaryotic extract cell-free expression system is the most popular and is commercially available [9]. We will mainly discuss the *E. coli* cell-extract system, termed as 'TXTL' in this section [10]. Compared to in vivo systems, the cell-free TXTL platform enables rapid prototyping of genetic circuit design using either generic plasmid DNA templates or short linear DNA templates [11,12]. Further, because TXTL-based circuits are implemented in vitro, these circuits are not limited by production of toxic proteins and chemicals or use of unnatural amino acids, which limit implementation of the same circuits in living cells [13,14].

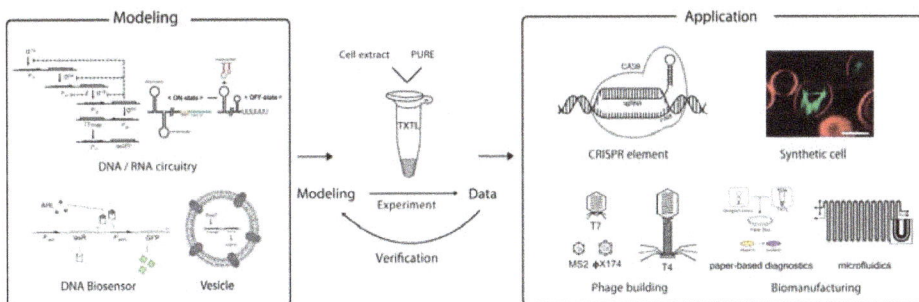

Figure 1. Overview of the cell-free transcription–translation platform. The cell-free transcription–translation platform including *Escherichia coli* cell-extract (TXTL) system and PURE system, allows for the prototyping of synthetic circuits rapidly through iterative cycles of experiments and computational modeling. TXTL has a number of applications, such as characterization of CRISPR elements or construction of synthetic cells. Reproduced with permission from [15,16].

The TXTL platform is not without limitations and challenges. Energy sources can be easily depleted in batch mode [17], while enzymes can degrade nucleic acids and protein products within the cell extract. Additionally, a complete understanding of machinery in TXTL system has yet to be achieved, and the yields of TXTL systems can be less than yields of corresponding in vivo systems [6]. Molecular crowding effects [18] or unintentional crosstalk between components [19] could contribute to these issues.

The PURE system is a completely purified cell-free expression platform containing the T7 RNA polymerase (RNAP) with fewer active components than cell extract-based TXTL systems [20]. In principle, the concentration of individual components can be adjusted in the PURE system during reconstitution, and purification of output proteins is straightforward by using affinity chromatography. The PURE system is costly and typically has a smaller yield than TXTL, but it can be advantageous for applications that require clear background and long-term storage of genetic elements [21].

The unique advantages of cell-free reactions make TXTL an ideal platform for prototyping genetic circuits by characterizing the properties and activities of circuit components [22]. For instance, the behaviors of CRISPR components (gRNA, protospacer adjacent motif (PAM) sequence, Cas9, and inhibitors of Cas9) are characterized using TXTL [23]. Importantly, the design–build–test cycle can be performed much faster than in vivo systems, thereby facilitating rapid prototyping of engineered circuits [24–26]. Circuit elements characterized in TXTL can be ported to an in vivo system, as demonstrated by the three- and five-node oscillator systems characterized in TXTL and successfully ported to *E. coli* [27].

Early works by Noireaux and colleagues demonstrated a multistage cascade circuit by superimposing several basic (input–parameter–output) units [11], where bacteriophage RNAP drove the expression of cascade circuits. However, with cascades including three to five stages, circuit performance was limited because of simple regulatory structures and extensive resource utilization. To expand the repertoire of transcriptional regulatory elements, sigma factors and cognate *E. coli* promoters were used for circuit construction [10,22]. They were able to demonstrate a five-stage transcriptional activation cascade through clever use of the different affinities of sigma factors to the core RNAP for efficient signal propagation (Figure 2A, left). Regulatory functions can be expanded by integrating various regulatory elements for constructing a more complex circuit (Figure 2A, right). Simultaneously monitoring the concentrations of produced RNA and proteins can assist in debugging synthetic circuits characterized in TXTL [19].

Synthetic RNA circuits are also efficiently and easily characterized in TXTL. Networks constructed from riboregulators propagate signals directly as RNAs, thus bypassing intermediate proteins, making these networks potentially simpler to design and implement than transcription factor-based layered circuits [28]. qPCR and next-generation sequencing techniques can characterize species, structural states, and interactions of RNAs [29]. Since the speed of signal propagation within circuits is determined by the decay rate of the signal, RNA networks can operate on much faster time scales than protein networks [30]. An early model of an RNA circuit used the transcriptional attenuator structure of RNA and its complementary antisense RNA [28]. The hairpin structure of the transcriptional attenuator was targeted by antisense RNA, which promoted the formation of a downstream intrinsic terminator hairpin that caused RNAP to fall off and stop transcription (Figure 2B). Other simple RNA-based circuits have also been characterized in TXTL systems [26,31], such as a negative autoregulation circuits, which use the attenuator and antisense RNA simultaneously [32].

The strength of the TXTL platform enables the expression of remarkably large natural DNA programs. A large amplification of the T7 phage, with a genome size of 40 kbp and supplemented with thioredoxin, was observed in vitro [33]. Cell-free self-organization of the even larger T4 phage, with a genome size of 169 kbp, under in vitro conditions was observed by increasing molecular crowding effects [34]. Replication of viral genomes occurred simultaneously with phage gene expression, protein synthesis, and viral assembly.

Beyond scientific inquiry, several practical tools emerged for using cell-free expression platforms. For instance, sequence-specific colorimetric detection of Zika viral RNA can be performed at single-base resolutions through a cell-free reaction on a paper disc. This paper-based diagnostic platform is advantageous because it is mobile and low-cost [35]. Another recent development demonstrated microfluidic reactors [15] and paper-based devices [36] that produced therapeutic proteins on demand.

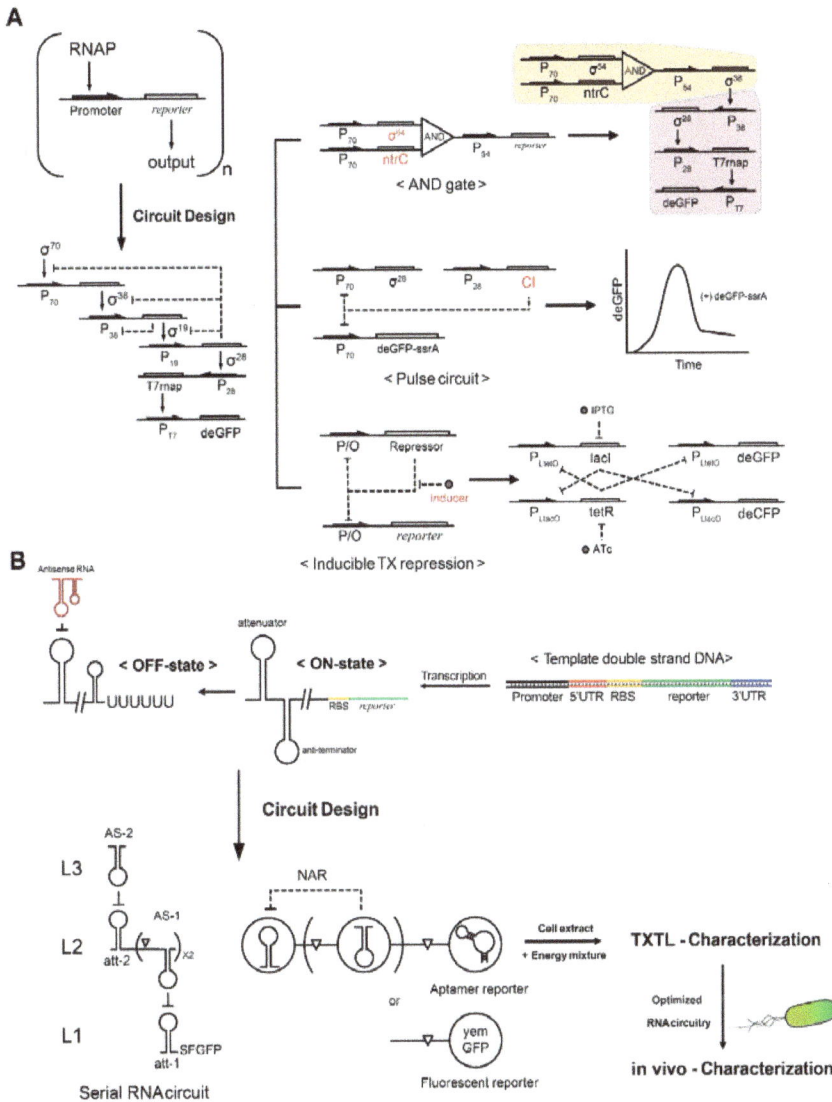

Figure 2. Systematic construction of DNA and RNA circuitry in TXTL. (**A**) Basic (input–parameter–output) modules are integrated to build complex synthetic circuits. Assembly of an AND gate, repressor, and inducer modules provides versatile and scalable circuits for synthetic biology applications. (**B**) RNA regulatory motifs are utilized for synthetic circuits such as a serial RNA circuit and a negative autoregulation circuit (NAR). The RNA circuits can be optimized in TXTL and ported to in vivo conditions.

3. In Vitro Synthetic Gene Circuits

In vitro regulatory networks are model systems that offer a flexible test bed for the design principles of biochemical networks without the complexity of cellular environments. In vitro regulatory models can be stripped of cellular machinery for protein translation and may use nucleic acid-based programs to design biochemical networks. In this section we will discuss simplified in vitro synthetic regulatory models using the synthetic transcription-based genelet system as an example [37].

Genelets are synthetic DNA switches that form a partially double-stranded (ds) DNA template. The expression states of a genelet are controlled by specific DNA inputs, which are recognized by an incomplete promoter region in the template [38–40]. The genelet system consists of synthetic DNA templates and two enzymes: T7 RNAP and *E. coli* ribonuclease H (RNase H). Because of the incomplete, partially single-stranded (ss) promoter region of genelets, the DNA template ('T') by itself is poorly transcribed [41]. An ssDNA activator ('A') can bind to complete the promoter region, and the resulting complex ('T-A') transcribes well, approximately half as efficiently as a full dsDNA template [39]. The activity of genelets can be controlled by nucleic acid inputs forming an inhibitory regulation [39] and an excitatory regulation [42]. The inhibitable switch is turned off by an RNA inhibitor that binds to DNA activator more favorably than the switch template thereby removing the activator from the template (Figure 3A). The activating switch is turned on by an RNA activator that binds to a DNA inhibitor and releases the DNA activator (Figure 3B). Both the DNA inhibitor and activator contain a 'toehold', a single-stranded overhang, to facilitate toehold-mediated strand displacement reactions [43]. Genelet circuits have the advantages of modularity and programmability for switch parameters, such as concentrations of switches and activators, which are analogues of weights and thresholds of neurons in artificial neural networks [38].

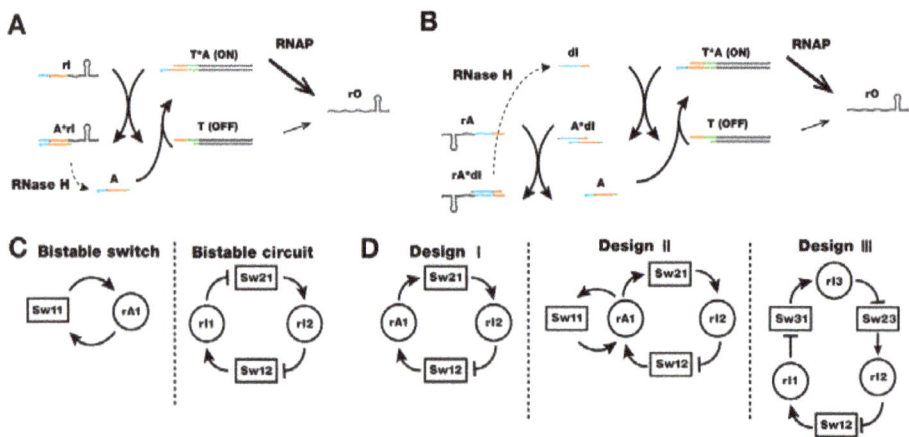

Figure 3. Genelet switches and circuits. (**A**) Design and operation mechanism of an inhibitable switch. The input, RNA inhibitor, sequesters the DNA activator from the active template and turns the switch to an OFF state. (**B**) Design and operation mechanism of an activating switch. The input, RNA activator, strips off DNA inhibitor bound to DNA activator. The released DNA activator in turn can turn the switch to an ON state. The sequence domains are color coded to indicate identical or complementary sequences. (**C**) Schematics of bistable circuits. A single switch with positive autoregulation (left) or two mutually inhibiting switches (right) can show bistability. (**D**) Schematics of oscillators. An activating switch and an inhibiting switch (Design I), Design I with further positive-autoregulation (Design II), and three inhibiting switches in a ring (Design III) form an overall negative feedback to achieve oscillation. Reproduced with permission from [15,16].

A bistable network is a dynamic system with two distinct stable equilibrium states [44], and it is often found in cellular networks requiring decision making processes such as cell cycle regulation, cellular differentiation [45], and apoptosis. The bistable network can be designed by genelets in two ways [39,42]: two switches can be connected in a mutually inhibiting configuration, or a single switch can be connected to activate its own transcription (Figure 3C). An oscillator circuit that produces periodic signals is another hallmark of basic circuit elements, and it is often found in cell signaling systems including genetic oscillation [46,47]. A synthetic oscillator was constructed using genelets with three different designs [48] (Figure 3D): a two-switch negative feedback oscillator that utilized activating and inhibiting connections (Design I); an amplified negative feedback oscillator that included an additional self-activating connection (Design II); and a three-switch ring oscillator with three inhibitory connections (Design III). The three designs shared the same basic architecture of overall negative feedback in the system. An amplified negative feedback oscillator (Design II) could potentially have four different phases, unlike a simpler oscillator (Design I), and the ring oscillator with an extra connection (Design III) featured a slower oscillation. The ability to construct different circuit motifs using genelets demonstrated the modularity and programmability of the system design. However, it remains a challenge to maintain circuit operation, such as oscillation, for an extended period of time in batch mode because of the exhaustion of nucleoside triphosphate (NTP) fuel, loss of enzyme activities, and build-up of incomplete RNA transcripts [48].

In addition to providing basic motifs, these synthetic circuits could be coupled with downstream processes to dynamically control other molecular systems. A number of downstream processes, which can be considered as a downstream 'load', have been demonstrated including DNA-based nanomechanical devices ("DNA tweezers") and functional RNA molecules ("aptamers") [40] (Figure 4A). DNA tweezers have two rigid double-stranded "arms" connected by a single-stranded hinge, which can be opened and closed by nucleic acid inputs. Retroactivity of the load process degraded the upstream oscillator circuit, which was alleviated by introducing insulator circuits to prevent excessive consumption of core oscillator components and to amplify RNA signals.

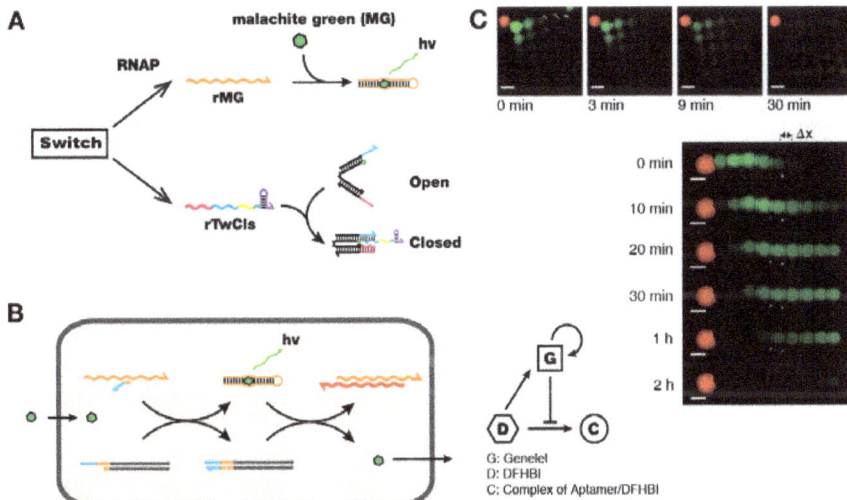

Figure 4. Extension of genelet circuits. (**A**) Driving downstream processes with genelet circuits. The output signal from genelet circuits can be functional RNA aptamers or can be used to drive DNA nanodevices such as DNA tweezers. (**B**) Signal propagation using encapsulated genelet circuits. Each droplet contains a genelet switch and aptamer-activator complex. Signal molecule (DFHBI) binds the aptamer and releases the DNA activator for the genelet switch. The activated genelet in turn produces kleptamer that binds the aptamer, releases DFHBI, and attenuates fluorescence output. (**C**) The experimental fluorescence images for one- and two-dimensional signal propagation. Reproduced with permission from [15,16].

Aptamers are nucleic acid molecules that fold into complex 3D shapes and bind to specific targets [49]. Functional RNA aptamers can be generated in vitro and tailored for a specific target, which are attractive features as downstream components to be controlled by genelet circuits. For instance, the transcription process can be monitored by using the aptamer against chromophore malachite green (MG) [40]. Sensing of specific molecules is enabled by designing the activator sequence of a genelet switch to bind to a specific aptamer, where the recognition of analyte by its aptamer releases the previously occupied DNA activator to activate the genelet. Using this approach, Dupin and Simmel demonstrated a genelet system to detect signal propagation in one and two dimensional array of compartments [50] (Figure 4B,C). The activity of enzymes can be controlled by using aptamers against T7 and SP6 RNAP, and an ssRNA/ssDNA with the complementary sequence of the aptamer, termed kleptamer, to provide yet another building block for synthetic biological circuitry with genelets [51]. These RNAP aptamers can also be used for logic circuits and transcriptional cascades [52].

Systems from natural processes and engineering disciplines provide further directions for developing genelet systems. Inspired by the architecture of electronic flip-flops, a genelet system design was proposed where the periods of a molecular clock were multiplied and divided [53]. Negative autoregulation provided model circuitry to produce outputs suitable for variable demands [54]. Adaptation in biological systems provided a framework to develop genelet circuits that detected fold-change of inputs [55]. Further, molecular titration utilized in natural and synthetic biological circuitry could be reiterated in genelet circuits with the support of mathematical modeling [56].

An analogous system, termed RTRACS (reverse transcription and transcription-based autonomous computing system), that relied on reverse transcriptase, DNA polymerase, RNAP, and RNase demonstrated modularity and programmability [57,58]. The modules of RTRACS received specific RNA input sequences and produced an RNA output through programmed computation. Experimental operation of an AND gate was demonstrated with RTRACS, and the prospect of more complex functionality such as oscillations was reported. The polymerase exonuclease nickase (PEN) toolbox bypassed the transcription step and relied exclusively on DNA and DNA-modifying enzymes to construct desired circuits [59,60]. Single-stranded DNA templates served as network architecture and short ssDNA species took the role of dynamic species that functioned as activators and inhibitors of templates. Despite its simplicity, the PEN toolbox successfully demonstrated bistability [60,61], oscillations [59,60,62], and pattern formations [63] through rational design approaches and easy monitoring [64]. An even more abstract approach is feasible with precisely programmed DNA sequences. Numerous studies demonstrated the power of DNA strand displacement circuits, including instructions, to create chemical reaction networks [65], logic circuits [66,67], neural networks [68,69], and oscillators [70] through toehold-mediated strand displacement [43]. These theoretical and experimental developments will enable future works to further enhance the programmability and complexity of synthetic in vitro circuits to control nucleic acid nanorobots for in vivo applications [71].

4. RNA Regulatory Circuits for Cell-Free Synthetic Biology

The programmable nature of RNA molecules that allows predictable design of structure and function provides a rationale to construct synthetic biological circuits with RNA toolkits. The most basic regulatory mechanism of RNA is to induce a trans interaction between the target mRNA and complementary RNA; RNAs that perform this function are called riboregulators [72]. Inspired by a plethora of natural examples of riboregulator-based gene expression control [73], synthetic biologists harnessed these design principles to create synthetic riboregulators in *E. coli* [74]. Following these seminal synthetic riboregulator systems, RNA-based synthetic biological circuits have emerged that are easily programmable with improved performance. In this section, we will discuss the recent progress in synthetic RNA regulators for cell-free diagnostic applications using toehold switch and small transcription activating RNA (STAR) as examples.

A toehold switch is a de novo-designed regulatory RNA inspired by the mechanism of a conventional engineered riboregulator [75] (Figure 5A). In the switch RNA, the ribosome binding site (RBS) and the start codon are blocked by its own secondary structure. When the trigger RNA is introduced to initiate a toehold-mediated branch migration, a switch-trigger complex is formed in which the RBS and the start codon become available for the expression of the target gene. In *E. coli*, high-performance toehold switches showed dynamic ranges rivaling those of well-established protein regulators. This suggests toehold switches can provide a novel, high-performance platform for synthetic biological circuits. Moreover, the RBS and the start codon are not directly involved in base-pairing within the secondary structure design of switch RNA, which allows for the construction of a library of toehold switches without sequence constraints.

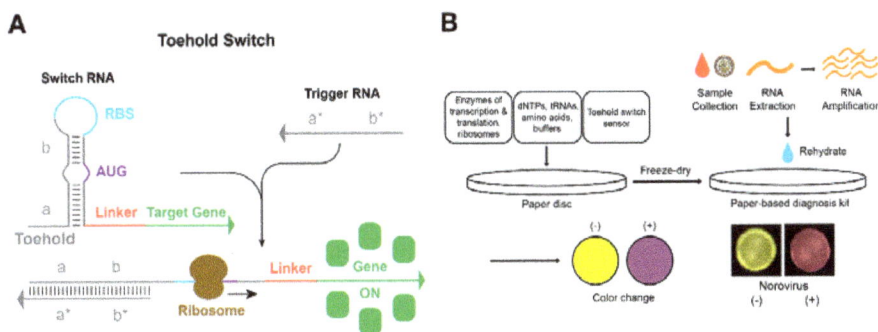

Figure 5. Toehold switch mechanism and application for paper-based diagnostics. (**A**) Mechanism of toehold switch. Linear-linear interaction between switch RNA and trigger RNA initiates from the toehold region. The resulting conformation of switch-trigger complex makes ribosome binding site (RBS) and start codon (AUG) available for ribosome access. (**B**) Freeze-dried paper-based diagnostic kit using toehold switch as a synthetic sensor. LacZ was used as a reporter gene so that the change of color could be checked by the naked eye.

Capitalizing the functionality of toehold switches, Pardee et al. constructed a paper-based diagnostic platform using a toehold switch as a sensor [21] (Figure 5B). DNA that encoded the switch RNA and components for cell-free expression (enzymes, dNTPs, amino acids, etc.) were freeze-dried on paper discs, which could remain stable for storage at room temperature. Upon the addition of the trigger RNA specific to toehold switches, up to 350-fold induction of green fluorescent protein (GFP) was observed with the desired orthogonality. This laid the foundation for development of a paper-based diagnostic tool for Ebola virus by using toehold switches that specifically sensed part of nucleoprotein mRNA of Ebola as trigger RNAs [21]. β-galactosidase (LacZ) was used as a reporter gene to allow confirmation of results with the naked eye, and 24 toehold switches that targeted different sequences (Sudan strain 12 regions, Zaire strain 12 regions) were successfully tested. One notable feature was that the fold change of LacZ expression was dependent on the sequence of switch RNA, suggesting that the sequence design needed to be optimized for improved utility. In a follow-up study, Pardee et al. constructed a paper-based diagnostic tool for Zika viruses [35]. To improve sensitivity, a nucleic acid sequence-based amplification (NASBA) step was introduced to isothermally amplify the target region of Zika RNA. Their system responded normally to Zika virus but not to the closely related Dengue virus. In addition, they combined this system with a CRISPR/Cas9-based module to create a NASBA-CRISPR cleavage (NASBACC) system. This biosensor showed sophisticated diagnostic performance that could discriminate strains of Zika viruses (American or African) by utilizing the presence of the PAM sequence. In another recent work, Ma et al. demonstrated a paper-based cell-free diagnostic system that detected Norovirus with toehold switches [76] (Figure 5B). They introduced virus enrichment via synbody and α-complementation of LacZ enzyme to improve the sensitivity and speed of diagnosis. Takahashi et al. demonstrated a microbiota sensing system, rather than a single virus, with the same platform [77]. They designed switch RNAs based on the 16S rRNA sequence of each bacterial species, and functionality of the sensor was verified against 10 different bacterial strains. Moreover, they proposed the potential for paper-based diagnosis of more diverse target RNAs, including host biomarker mRNAs such as calprotectin, Interleukin 8, C-X-C motif chemokine ligand

5, oncostatin M, and specific pathogen toxin mRNA. These demonstrations provide evidence that toehold switches can be utilized as a generalizable platform for portable diagnostic systems in field testing diseases and other environmental samples.

To process an increasing complexity of inputs through synthetic biological circuitry, it is necessary to integrate a number of input signals in a seamless manner. To achieve this goal, the basic mechanism of the toehold switch has been expanded to incorporate multiple toehold switches in the same RNA transcript, which facilitates signal integration, termed a 'ribocomputing' strategy [78]. Green et al. implemented a complex logic system (combination of AND/OR/NOT) of 12 RNA inputs with five consecutive toehold switches colocalized in the same transcript. This 12-input logic circuit in *E. coli* provided evidence that a ribocomputing strategy could help in scaling up synthetic biological circuits in the future [79]. At the same time, novel RNA tools where translation is inactivated by trigger RNAs are also being explored. These include toehold repressor, three-way junction (3WJ) repressor, and looped antisense oligonucleotide, which enable a more complex and versatile logic with universal NAND and NOR gates [80,81].

In a similar vein, Lucks et al. engineered the natural antisense RNA-mediated transcriptional attenuation mechanism of plasmid pT181, and they proposed an RNA toolkit that could turn off transcription [28]. Based on this, Takahashi and Lucks demonstrated that more diverse orthogonal transcriptional regulators could be designed by combining the module with natural antisense RNA regulators [82]. Building on these works, Chappell et al. devised a novel transcription regulatory system, termed the 'small transcription activating RNA' (STAR), which promoted transcription upon cognate trigger RNA binding [83] (Figure 6A). The natural mechanism was utilized in an opposite manner such that the complementary STAR RNA disrupted the transcription terminator structure of the target gene. In their first demonstration of the STAR system, the fold change ranged from 3- to 94-fold. In a subsequent work, they further optimized various domain lengths of STAR RNA through computational designs to create a STAR library with a broad fold activation range, from more than 400-fold to less than 10-fold, to allow for more sophisticated biological circuit designs [84].

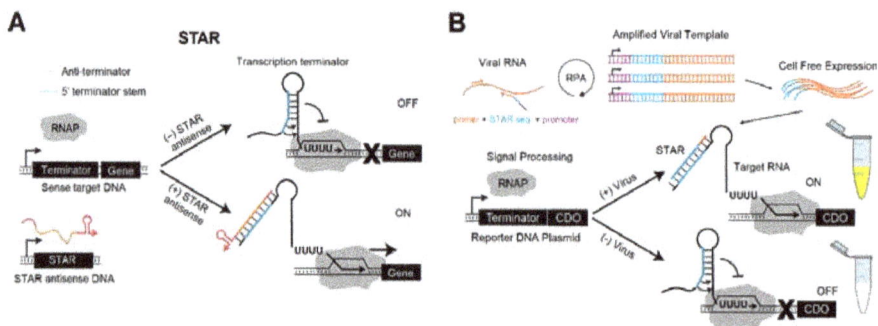

Figure 6. Small transcription activating RNA (STAR) system and application for detecting plant pathogens. (**A**) Mechanism of STAR. A transcription terminator consists of a stem-loop and a poly-U track, where the binding of STAR RNA breaks the step-loop structure such that transcription proceeds normally. (**B**) A platform to diagnose plant pathogens using STAR. Viral RNA in the sample can be amplified with the addition of a STAR sequence and a promoter through recombinase polymerase amplification (RPA). Then, the corresponding RNA transcribed through cell-free expression induces the expression of reporter gene (CDO, catechol 2,3-dioxygenase). RNAP: T7 RNA polymerase.

A platform for plant pathogen detection was demonstrated using the STAR system [85] (Figure 6B). Verosloff et al. amplified viral DNA with the T7 RNAP promoter and an upstream STAR sequence by using recombinase polymerase amplification (RPA) with a primer that bound to a specific sequence of viral DNA. When viral RNA with the STAR sequence was transcribed by cell-free expression, the reporter RNA started normal transcription of catechol 2,3-dioxygenase (CDO) in the tube, which caused a colorimetric change in the sample observable with the naked eye. In particular, this system has the advantage that both the RPA and the cell-free expression steps are conducted isothermally using only body heat as a heat source.

A number of other works also demonstrated the utility of synthetic RNA regulatory parts for building synthetic biological circuits. Well-established anti-sense RNA (asRNA) could be utilized for translation regulation through binding at the 5′ untranslated region (UTR) and the start codon of the target mRNA [86]. Through analysis of the difference in repression of asRNA sequences, the repression was further enhanced by introducing the Hfq binding sequence into the asRNA [87]. Meanwhile, Rodrigo and Jaramillo developed a computational design tool, named 'AutoBioCAD', that allowed automatic RNA circuit design using secondary structure design and free energy analysis [88,89]. These RNA regulatory toolkits can contribute to the growing repertoire of cell-free synthetic biology applications including point-of-care devices for biomedical applications.

5. Encapsulation of in Vitro Circuits toward the Synthesis of Artificial Cells

In vitro synthetic biology has recently made progress towards realizing minimal cell systems. A key step toward this is the encapsulation of gene expression or TXTL systems in microscopic compartments. It has been demonstrated that both the kinetics and noise levels of chemical reactions are different in bulk than in cell-sized compartments or molecularly crowded solutions [90,91]. Thus, working with encapsulated synthetic biology components may improve our understanding of native cellular systems and how to emulate cell processes in synthetic systems [16]. Minimal cell systems can be designed to perform specified tasks autonomously, or they can network with native cells to increase sensing and actuation of biological systems. As encapsulation offers a barrier between critical components in synthetic circuits and surrounding environments, it is an essential step for designing effective minimal cell systems.

Both water-in-oil droplets and vesicles have been used to encapsulate gene expression systems in sizes relevant to cells (roughly 1–50 μm in diameter). Water-in-oil droplets have been demonstrated to be biocompatible, stable at high temperatures, and capable of withstanding deformation [92]. The oil medium surrounding the droplets greatly limits molecular exchange between individual droplets, effectively creating isolated, independent, cell-sized reaction chambers. Microfluidics can produce hundreds or thousands of uniformly sized droplets a minute, while shaken droplet protocols result in droplets of varying sizes.

Liposomes or vesicles more closely resemble native cells than water-in-oil droplets, but they are non-trivial to produce at cell sizes. Because the environment surrounding the vesicles is aqueous, exchange of biological molecules between the vesicles and the environment is possible for membrane-permeable molecules as well as membrane-impermeable molecules in the presence of surface pores or channels [93–95]. Emulsion transfer, thin-film hydration, and microfluidics have been shown to effectively encapsulate TXTL systems in vesicles [90,96–100]. Emulsion transfer and microfluidics allow for finer control of the vesicle size and contents than thin-film hydration techniques [100].

Early encapsulation of expression systems characterized the production of single reporter proteins in bulk and in vesicles. Noireaux and Libchaber reported that expression of GFP in both vesicles and bulk solution had similar durations and produced similar outputs [90] (Figure 7A). They showed that by expressing a GFP-labeled α-hemolysin pore, expression was increased by one order of magnitude. The toxin α-hemolysin acts as a pore in lipid bilayers with a molecular mass cutoff of 3 kDa, which allows nutrients from a surrounding feeding solution to enter the vesicle. Tan et al. demonstrated that vesicles protected an encapsulated GFP expression system from RNase A in the aqueous medium surrounding the droplet, which inhibited gene expression via degrading RNA [101].

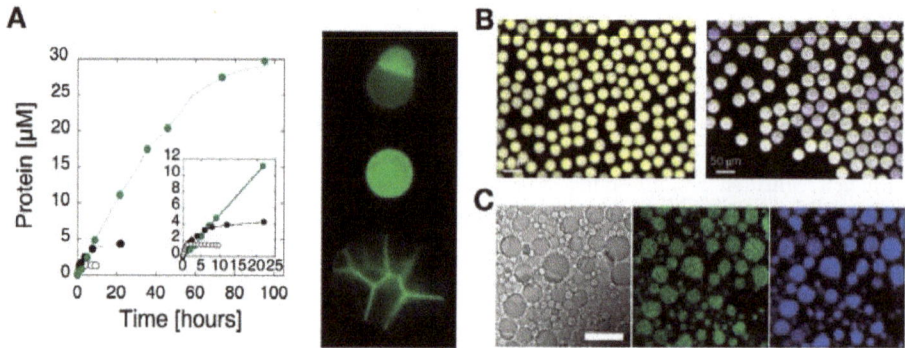

Figure 7. Synthetic gene circuits in cell-sized compartments. (**A**) (left) Expression of eGFP in bulk (open circles) and in a vesicle (closed dark circles), and expression of α-hemolysin-eGFP (closed green circles). (Inset) an expanded view of the first 20 h. (right) Fluorescence microscopy images of α-hemolysin-eGFP expressed in vesicles. Scale bar: 10 μm. Reproduced with permission from [16]. (**B**) Superimposed false-color images of cyan fluorescent protein (CFP) and yellow florescent protein (YFP) expressed in droplets without (left) or with (right) Ficoll. In the presence of Ficoll, the expression level is highly variable across the population of droplets. Reproduced with permission from [102]. (**C**) Microscopy images of mYPet expression inside an aqueous two-phase system (ATPS) water-in-oil droplets. The images of transmitted light, fluorescence microscopy of mYPet, and fluorescence microscopy of Alexa 647-labeled dextran are presented, from left to right. mYPet is preferentially expressed in the dextran phase rather than the polyethylene glycol (PEG) phase. Scale bar: 25 μm. Reproduced with permission from [103].

Actualization of more complex gene expression systems, such as oscillators and cascading gene circuits, have subsequently been described in compartments [97,104]. In cascading reactions, products of an initial transcription system are necessary for further TXTL processes in the circuit. Garamella et al. encapsulated both five- and six-gene cascading circuits within vesicles [22]. They reported both an increase in the average expression across a population of 20–30 vesicles as well as an increase in the variability of expression for individual vesicles within the population for the six-gene circuit compared to the five-gene circuit. Adamala et al. reported that higher-order cascading circuits in vesicles produced similar amounts of protein to bulk reactions containing the same volume [105]. They observed smaller vesicles than the work by Garamella and colleagues, using mammalian HeLa cell extracts, and their circuits were triggered by diffusion of doxycycline through α-hemolysin pores in the membranes of the vesicles. The average expression for the population of vesicles containing a three-gene cascading circuit produced less fLuc than vesicles containing either one- or two-gene cascading circuits. The encapsulated three-gene circuit did, however, produce similar amounts of fLuc to a bulk solution with the same reaction volume, while the encapsulated one- and two-gene circuits produced less than the corresponding bulk reactions. Both reports noted that the high variability in expression for the higher-order circuits likely was due to nonuniform encapsulation of the individual components of the circuit throughout the population of vesicles. This phenomenon has been described in other works, and it even affects lower-order genetic circuits such as simple transcription of eGFP [106].

In addition to studying the effects of compartmentalization on TXTL systems, the influence of molecular crowding on encapsulated systems is relevant for both minimal cell design and understanding native cell processes. The crowded interiors of cells have been shown to influence intracellular reaction rates [107]. Hansen et al. investigated the relationship among stochasticity of expression within water-in-oil droplets, concentration of genes, and concentration of crowding molecules [102]. They reported that introduction of the crowding molecule Ficoll 70 resulted in microenvironments of cyan fluorescent protein (CFP) and yellow fluorescent protein (YFP) within the droplets for the duration of expression (Figure 7B). They concluded that the microenvironments formed because the rate of mRNA production was greater than the diffusion rate of the macromolecules involved in TXTL, as CFP and YFP diffused evenly through the droplets after the expression completed. They also noted that decreasing the available copies of gene within the expression system further increased the stochasticity of expression across the population of droplets. Tan et al. reported the effects of crowding molecules on minimal gene expression systems in vesicles, noting that crowding due to large dextran polymers increased expression of GFP in larger vesicles, while it had little effect on expression in smaller vesicles [101].

Molecular crowding not only affects reaction rates within cell-sized compartments but also leads to crowding-induced phase separation, which is another area of interest for synthetic biology efforts [108,109]. Liquid phase separation is a form of membrane-less partitioning, which occurs in native cell structures such as nucleoli and centrosomes, and is influenced by temperature, pH, concentration, and other factors [110]. Torre et al. showed confined expression of mYPet, a fluorescent protein, within a single phase of an aqueous two-phase system (ATPS) [103] (Figure 7C). The ATPS was a result of introducing both polyethylene glycol (PEG) and dextran to water-in-oil droplets, which separated into distinct PEG- and dextran-rich phases within the droplet. Torre et al. hypothesized that confinement of the gene expression to the dextran phases was a result of TXTL machinery partitioning the less hydrophobic dextran phase of the droplet. They reported no expression in encapsulated aqueous three-phase systems, suggesting this was due to splitting of the TXTL components between the dextran and Ficoll phases within the droplets.

As the complexity of artificial cells increased, so too has the exploration of communication between networks made of artificial and native cells [50,111–114]. After showing that two distinct liposome-based minimal cells in a shared environment could respond to the same trigger without crosstalk, Adamala et al. demonstrated cascading networks of synthetic minimal cells as well as fusion-controlled TXTL systems [105] (Figure 8A). They realized a cascading expression system in which products from one vesicle that contained bacterial TX machinery triggered a response in a separate vesicle that contained mammalian TL machinery. Their work demonstrated that encapsulating TXTL systems provided modularity, allowing for interaction between otherwise incompatible components. Lentini et al. increased the sensing capacity of *E. coli* through networking with a synthetic translator cell [115] (Figure 8B). They induced expression of GFP within *E. coli* by creating an artificial cell that produced a chemical signal familiar to the bacteria, isopropyl β-D-1-thiogalactopyranoside (IPTG), in response to theophylline (Theo), which was otherwise undetectable to the native cell.

The exploration of artificial TXTL systems in cell-sized compartments is necessary for realizing artificial cells and fully understanding intracellular processes in confined and crowded environments. A number of complex gene expression circuits have been demonstrated in cell-sized compartments; however, the variety of components available for design—lipid bilayer vesicles vs water-in-oil droplets, bacterial vs mammalian cell-extracts, and so on—makes direct comparison between different studies difficult. The often encountered high variability in component concentrations during encapsulation processes may be alleviated through further exploration of encapsulation methods, fusion, and intercompartment exchange processes. Through compartmentalization, previously incompatible natural or synthetic systems in bulk solution can be interconnected as modular components, which paves the way for increasing the complexity of cell-free synthetic circuits and coordination with native cells.

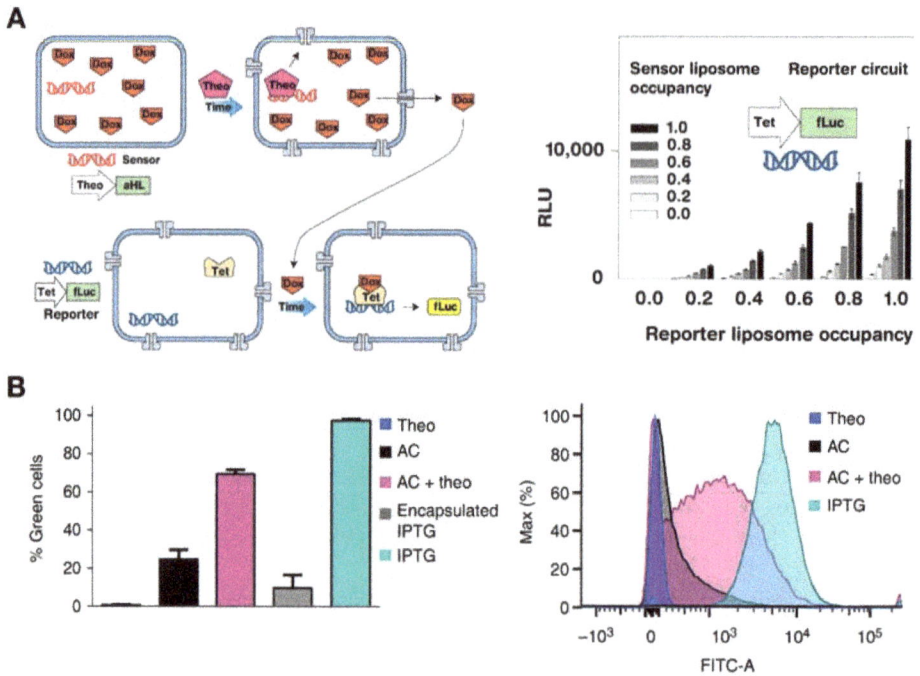

Figure 8. Interconnecting artificial and natural cells. (**A**) (left) Schematic design of synthetic sensor and reporter liposome pair, which contain bacterial and mammalian TXTL machinery, respectively. α-hemolysin (aHL) produced by theophylline treatment in the sensor liposome releases internal doxycycline to the environment, which in turn triggers expression of fLuc in the reporter liposome. (right) Expression of fLuc in different ratios of sensor and reporter liposomes. Occupancy refers to the ratio of droplets that contain TXTL machinery for both sensor and reporter droplets. Reproduced with permission from [105]. (**B**) (left) Flow cytometry data for *E. coli* containing a plasmid for GFP in the presence of the following components: theophylline (Theo), artificial cells (AC), artificial cells with theophylline (AC + Theo), isopropyl β-D-1-thiogalactopyranoside (IPTG) encapsulated in vesicles (Encapsulated IPTG), and IPTG in the bulk solution (IPTG). (right) Histogram of flow cytometry data shown in the left panel. Fluorescent signal is only increased in the presence of artificial cells and theophylline. Reproduced with permission from [115].

6. Artificial DNA Structures and Systems for in Vitro Synthetic Biology

Because of its predictable self-assembly properties, DNA has been used to build versatile molecular machines and structures [116,117]. Complementary Watson–Crick base pairing between segments of synthetically designed DNA strands is utilized for the rational design of these static and dynamic devices, which can function autonomously by processing information obtained from its surroundings. DNA systems and circuits can present properties comparable to devices naturally found in the living cell, for example, it is possible to build DNA nanotubes with size and mechanical properties comparable to cytoskeletal filaments, DNA nanopores that dock on lipid bilayers with selective permeability, and transcriptional oscillators that could serve as clocks in synthetic cells [40,48,118,119]. DNA nanostructures could serve as physical components such as scaffolds, pores, and transport elements in artificial cells. Nucleic acid strand displacement reactions could be used to build sensors and signaling pathways [66,120–122]. Yet, these synthetic DNA systems may have difficulty in achieving desired structural integrity and functionality in the cellular environment since DNA nanostructures and networks have been typically characterized in buffer conditions very different from the complex environment of a cell. Therefore, it becomes a necessity to explore synthetic DNA systems in cell lysates, serum, and cell-free extracts to develop design rules for proper operation in complex cellular environments and to realize their full potential as programmable components for in vitro and in vivo synthetic biology.

The presence of cytoplasmic enzymes can affect the structural stability of synthetic DNA systems. Kuem et al. measured the half-life of tetrahedral DNA nanostructures (TDNs) in the presence of DNase I and found that the stability of TDNs was more than twice that of double-stranded DNA [123] (Figure 9A). Castro et al. incubated DNA origami structures in the presence of different nucleases—DNase I, T7 endonuclease I, T7 exonuclease, *E. coli* exonuclease I, lambda exonuclease, and MseI restriction endonuclease [124]. Only DNase I and T7 endonuclease I were found to degrade the test origami structure, where the DNA origami structure could withstand complete degradation for 2 h in the presence of DNase I in contrast to the duplex plasmid DNA that disappeared within 5 min (Figure 9B). The interconnectivity and dense packing of the DNA nanostructures rendered some resistance to degradation by nucleases.

Another cell-like medium in which to characterize synthetic DNA systems can be cell lysates—the mixtures containing cellular components created by breaking down the membranes of cells. Mei et al. tested the stability of DNA origamis in cell lysate and reported that single- and double-stranded nucleic acids could not be recovered, whereas DNA origami could be recovered after up to 12 h [125]. However, the physiological relevance of this particular study was damped by the fact that cell lysate used sodium dodecyl sulfate (SDS) and deoxycholic acid (DCA), which suppressed many cellular enzymes. Therefore, a more physiological cell lysate preparation should be used for better assessing synthetic DNA systems in cell-like media.

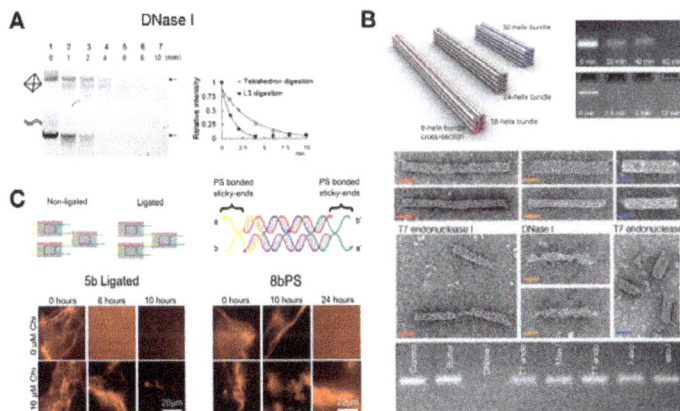

Figure 9. Characterization of DNA structures in cell-like media. (**A**) Denaturing polyacrylamide gel electrophoresis (PAGE) of tetrahedron and duplex DNAs with nonspecific degradation by DNase I. Digestion of the unligated tetrahedron is gradual and appears to generate a well-defined product, whereas digestion of linear DNA appears to be rapid and nonspecific. Reproduced with permission from [123]. (**B**) Stability of honeycomb-packed DNA nanostructure: 140 nm (18-helix bundle), 100 nm (24-helix bundle), and 70 nm (32-helix bundle), from left to right, were used for stability screening with TEM and agarose gel electrophoresis. Scale bar = 20 nm. Reproduced with permission from [124]. (**C**) Enhanced stability of DNA nanotubes with χ-site integration and chemical modifications in *E. coli* TXTL system. Fluorescence microscopy images of five-base DNA nanotubes with ligation of tile sticky ends and eight-base DNA nanotubes with phosphorothioate-bonded tile sticky ends incubated in TXTL with and without χ-site DNA present. Scale bar = 20 μm. Reproduced with permission from [126].

A useful platform for rapid characterization of synthetic components is the *E. coli* cell-free TXTL system. TXTL reiterates the physiological conditions found in cells as well as harsh linear DNA degradation through the RecBCD complex. Klocke et al. tested the stability of tile-based DNA nanostructures in the TXTL system, demonstrating that the stability of structures increased significantly in the presence of χ-site double-stranded DNA, which was an inhibitor of the RecBCD complex [126] (Figure 9C). With the addition of 10 μM χ-sequences, tile-based nanotubes assembled from ligated DNA strands were stable in TXTL for more than 10 h. Further, phosphorothioation of the strands within nanotubes extended their viability in TXTL for more than 10 h without, and 24 h with, χ-sequences. However, chemically modified strands in DNA structures can introduce toxicity or trigger unwanted immune responses when introduced in cells [127]. Thus, chemical modifications should consider potential trade-offs of structural stability and cell toxicity.

A number of studies were carried out to test the stability of DNA systems in serum and serum-supplemented media. The Sleiman group tested the stability of DNA assemblies in 10% fetal bovine serum (FBS) [128]. They reported that individual strands had a half-life of less than one hour, whereas the half-life of DNA structures in the shape of a triangular prism was closer to two hours. Hahn et al. tested the stability of DNA origamis in mammalian cell culture media supplemented with serum, and they indicated that DNA nanostructures were sensitive to depletion of Mg^{2+} in tissue culture medium [129] (Figure 10A). Interestingly, structural stability was significantly enhanced with the addition of actin, a protein that competitively binds to nucleases. No observable differences in cell growth, viability, or phenotype were present when actin was included in the medium.

The functionality of synthetic DNA circuitry is an important goal to achieve in the cellular environment. This spurred a number of studies on DNA circuitry in serum and serum-supplemented media. Goltry et al. investigated topological influences on the lifetimes of DNA devices using a three-state DNA tweezer nanomachine and a two-state linear probe in human serum and FBS [130] (Figure 10B). Degradation analysis revealed that the mean lifetimes of both systems in human serum were roughly six times longer than those in FBS. They reported that the device lifetimes varied greatly with topology (i.e., circular vs linear) and molecular conformation (i.e., shape of the structure), potentially providing a simple design rule to program structural stability or fragility. Graugnard et al. tested an autocatalytic strand-displacement network, reported by Zhang and colleagues [131], in human serum and mouse serum [132]. With the addition of SDS to halt nuclease activity, the synthetic network was functional in serum with both DNA and RNA catalysts. Fern and Schulman investigated strategies to enable strand-displacement circuits to operate in 10% FBS [133] (Figure 10C). By inhibiting nuclease activity using actin protein, and by modifying DNA complexes with hairpin extensions on the 3′ ends of DNA strands, the half-life of DNA strands increased by 10-fold. Through these modifications, a multilayer cascade circuit was demonstrated that released a desired output strand with controlled kinetics with the aid of computational modeling.

Taken together, densely packed and interconnected DNA nanostructures, such as DNA origami, are consistently more stable than structurally simple nucleic acid architectures in cell-like environments. Nucleases can be a primary cause for structural instability of synthetic DNA systems; however, other processes also need to be taken into consideration. For instance, nonspecific transcription by RNAP can produce transcripts that, in turn, can interact with DNA nanostructures, leading to disassembly via a toehold-mediated branch migration [134]. Thus, more systematic research is warranted to develop strategies that shield DNA systems from unintended crosstalk with biological components and that maintain integrity of devices within the cellular context. Use of actin or χ-sites as molecular decoys, structural modifications to increase interconnectivity, chemical modifications, and hairpin extensions are some of the strategies explored towards achieving better functionality of synthetic DNA systems. The improved design rules for DNA nanomachines and circuits may support the translation of devices operational in cell-free settings to the cellular environment.

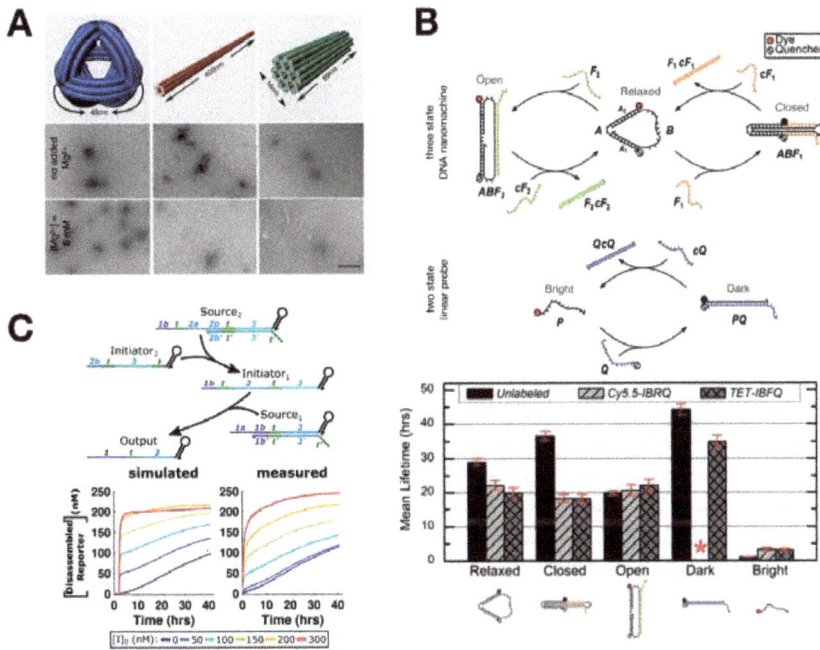

Figure 10. Characterization of DNA structures and circuits in serum. (**A**) 3D model of DNA nanooctahedron (DNO), six-helix bundle nanotube (NT), and 24-helix nanorod (NR) (top). TEM images of nanostructures incubated in unmodified (middle) or Mg^{2+}-adjusted (bottom) medium. Structural integrity is maintained for all three designs with additional Mg ion. Scale bar = 100 nm. Reproduced with permission from [129]. (**B**) (top) Three-state DNA nanomachine transitions between relaxed, closed, and open states with the addition of fuel strands and their complements. The two-state linear probe transitions between bright and dark states upon hybridization of the dye-labeled probe strand, P, and the quencher-labeled strand, Q. (bottom) Mean lifetimes of the DNA nanomachines and linear probes show considerable differences in degradation rates. Reproduced with permission from [130]. (**C**) (top) Schematic of two-layer DNA cascade reaction. (bottom) Simulation results of a two-layer cascade with 5 bp toeholds using the fitted parameters and experimental measurements. Reproduced with permission from [133].

7. Mathematical Modeling Supports the Development and Analysis of in Vitro Systems

Mathematical modeling has contributed to the success of synthetic biology since its inception [1,2]. Models are helpful to support the design of synthetic systems and to explain quantitatively observed phenomena, which may be otherwise difficult to understand, especially when they include feedback loops. Validated models are also useful to make predictions and guide experiments, making it possible to save time and costly reagents. Ordinary differential equations (ODEs) are one of the simplest approaches to build mathematical models to describe kinetic systems. ODE models are particularly well-suited to capture systems operating at high copy numbers, so they are an excellent choice for in vitro synthetic biology. Because in an in vitro setting it is usually possible to collect a large amount of kinetic data in which experimental conditions are varied systematically, ODEs can be easily fitted to the data and yield solid estimates of various parameters that govern the kinetics.

Many in vitro synthetic systems have been quantitatively modeled using ODEs that can be built systematically starting from a list of relevant chemical reactions. Transcriptional networks, for example, include synthetic genes (genelets), two enzymes, and mRNA species to create regulatory interconnections between genelets in a rational manner [39,48]. To formulate an ODE model that captures the kinetics of an inhibited genelet, the species to be considered are the template T (active and inactive), its DNA activator A, the RNA inhibitor rI, and RNAP and RNase H that control RNA production and degradation. RNA inhibitor is produced by a "source" template (S), whose concentration is constant. The active template TA produces an RNA output rO. The template and its activator are referred to as a "switch" (SW). The complete set of reactions associated with this system is shown in Figure 11A. Using the law of mass action, it is possible to write ODEs that describe the reaction kinetics. For example, the free template concentration T is converted to active template TA by binding to the activator A at rate constant k_{TA}; in turn, the active template TA is converted back to the free template T when it interacts with the inhibitor rI, which displaces the activator at rate k_{TAI}. As a consequence, we can immediately write the kinetics of the free template as:

$$\frac{dT}{dt} = -k_{TA}[T][A] + k_{TAI}[TA][rI]. \tag{1}$$

The ODEs for all other species can be derived with the same procedure. Because the total template concentration remains constant, then $[TA] = \left[T^{tot}\right] - [T]$, which means the model does not require a specific ODE for the kinetics of TA. For enzyme kinetics, it is possible to use the well-known Michaelis–Menten quasi-steady state approximation so that the available concentrations of RNAP and RNase H can be expressed with an analytical, static formula as a function of their substrate (Figure 11C). The complete ODE model is pictured in Figure 11D. This model can be fitted to kinetic data, and it reproduces the steady state input–output map of the inhibitable switch (the input is the concentration of source S, the output is the fraction of active switch) (Figure 11E).

More complex transcriptional networks can be modeled with the same approach by modularly composing the models of individual switches. For instance, an ODE model of a bistable switch (Figure 11F) could be immediately built by interconnecting the models of two inhibitable switches whose RNA outputs mutually inhibited transcription [39]. The models were augmented by taking into account undesired or putative reactions, such as transcription from inactive template T, and captured very well the kinetic experiments, as shown in Figure 11G. Similarly, Kim and Winfree developed [48] ODE models for different versions of a transcriptional oscillator, in which side reactions played a very important role. For example, the ability of the model to reproduce the oscillator kinetics (in particular the damping rate) was significantly improved by including incomplete degradation products that accumulated during the oscillator reaction and their potential interactions with activation and inhibition of the genelets. ODE models built using the law of mass action can be used to model genelets, molecular machines, and other molecular processes, and they can be used to computationally test the influence of new, modified, or unknown components on the system [40,135]. To summarize, mechanistic ODE models are successful at recapitulating the dynamic behaviors of in vitro synthetic systems, and they can be easily expanded to include additional species or reactions. Yet, these models can become very large, even in systems with few desired interactions, and obtaining physically meaningful fitted estimates for the model parameters requires the inclusion of tight bounds.

A phenomenological approach to building ODE models is advantageous in building models with few variables and parameters, which helps in obtaining more intuitive results on the behavior of the system under consideration. Rather than being built from a list of chemical reactions, phenomenological models rely on qualitative relationships between species. For example, the steady state behavior of the inhibitable genelet shown in Figure 11E could be modeled using a single ODE in which the source template could cause a decrease of active switch via a Hill-type function. Beyond transcriptional circuits, phenomenological models have been used for many synthetic systems built in vitro, including RNA regulator-based circuits [32,136,137]. Figure 11H shows the qualitative model for a transcription regulator that achieves negative autoregulation (NAR) [32]; the species in the equations are the concentrations of RNA (R) and GFP (G), whose productions decrease as the concentration of RNA (R) increases (self-inhibition). Using parameters from the literature, this simple model was used to compare the efficiency of one versus two tandem repressors, and it yielded the trajectories in Figure 11I that qualitatively agreed with the experimental data in Figure 11J. Although a detailed mechanistic model was required to quantitatively reproduce data (Figure 11J), the simple model provided useful insights on the system kinetics.

Limited modeling efforts have been dedicated to compartmentalized cell-free circuits, largely because this research is still in its early stages. Encapsulation can introduce noise and stochastic phenomena even when operating with few components at high concentration. The operation of a transcriptional oscillator, for example, was significantly affected by partitioning noise when encapsulated in water-in-oil droplets [104]; a model combining ODEs and stochastic partitioning of components (following a Poisson process) was able to recapitulate the variability in the circuit dynamics. Stochastic simulations could improve our understanding of noise observed in recent works aimed at encapsulation in high-order synthetic circuits [105,111,138].

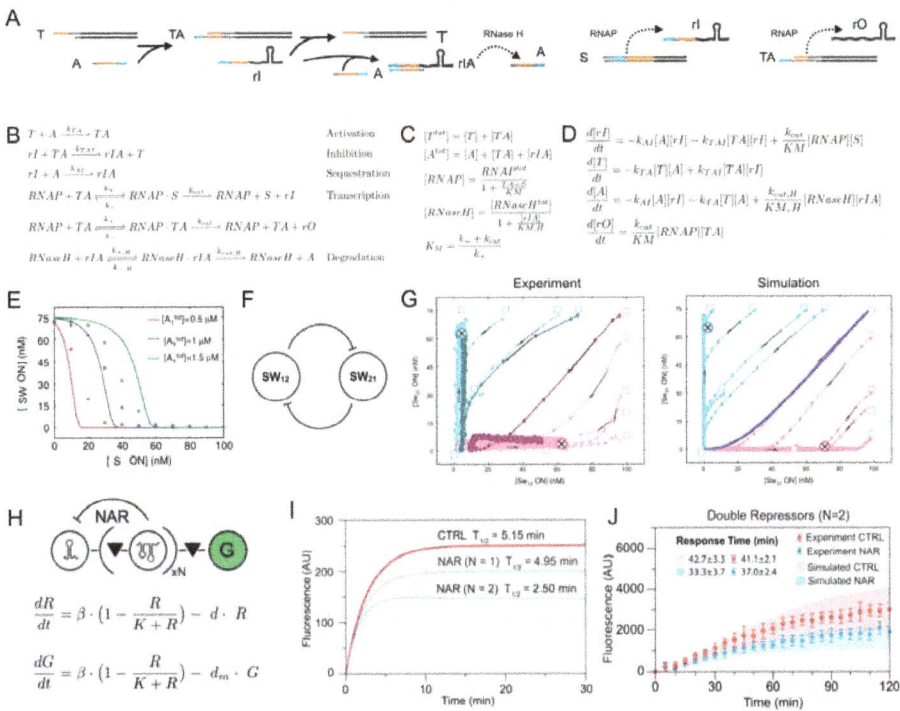

Figure 11. Ordinary differential equation (ODE) models for design and characterization of in vitro synthetic systems. (**A**) Schematic of an inhibited transcriptional switch. (**B**) List of reactions describing a transcriptional switch that is inhibited by RNA transcribed by a source template S. (**C**) Mass conservation and Michaelis–Menten expressions allow a simplification of the ODEs. (**D**) ODEs describing the inhibited transcriptional switch. (**E**) Example data showing the input–output steady state curve mapping the concentration of inhibitor source to the output switch concentration, overlapped with simulated steady state data (solid lines). Reproduced with permission from [39]. (**F**) A bistable switch can be constructed by interconnecting two inhibitor switches. (**G**) Experimental data (left) compared to simulated trajectories (right) of the bistable switch. Reproduced with permission from [39]. (**H**) Example transcription regulator used to build a gene performing NAR. G indicates GFP and its qualitative ODE model. (**I**) The qualitative ODE model suggests that the NAR circuit operates better when using at least two transcription repressors in tandem. (**J**) Simulations of detailed mechanistic models reproduce experimental data well. Adapted with permission from [32].

Methods Protoc. **2019**, *2*, 39

8. Concluding Remarks

As synthetic biological systems have become larger and more complex, deciphering the intricate interaction of synthetic systems and biological entities becomes a challenging task. Cell-free synthetic biological approaches, with the aid of rapid progress in its scope, and toolkits may provide the right platform for rapid design–build–test cycles. New technological breakthroughs for synthetic biology, such as CRISPR-Cas systems, can also be elucidated in this simplified TXTL test bed [23]. The ease with which to program nucleic acids has dramatically accelerated the structural and functional complexity of nucleic acid-based molecular devices. These new developments encompass simplified synthetic model dynamical systems and nucleic acid nanostructures, as well as synthetic RNA regulatory components, which form the core of practical tools for biomedical applications. Compartmentalization for synthetic cells opens up ways for scientific inquiry and enhanced functionality through networks of synthetic and natural systems. Data-driven model building needs to guide the research and development towards complex synthetic systems with prescribed dynamics in the future. In the coming years, we anticipate that the utility of cell-free synthetic biology will rapidly expand the scope of biotechnology and synthetic biology, and it will provide innovative solutions in biomanufacturing therapeutics for biomedical applications and biologic products for industrial applications.

Author Contributions: D.J., M.K., S.A., J.K., S.C., E.F., and J.K. summarized the data and wrote the paper.

Funding: This work was partially supported by the U.S. Department of Energy under grant SC0010595, which paid for the salary of M.K. and S.A.

Conflicts of Interest: The authors declare no conflict of interest.

References

1. Elowitz, M.B.; Leibler, S. A synthetic oscillatory network of transcriptional regulators. *Nature.* **2000**, *403*, 335–338. [CrossRef] [PubMed]
2. Gardner, T.S.; Cantor, C.R.; Collins, J.J. Construction of a genetic toggle switch in *Escherichia coli*. *Nature* **2000**, *403*, 339–342. [CrossRef] [PubMed]
3. Cameron, D.E.; Bashor, C.J.; Collins, J.J. A brief history of synthetic biology. *Nat. Rev. Microbiol.* **2014**, *12*, 381–390. [CrossRef] [PubMed]
4. Nielsen, A.A.; Der, B.S.; Shin, J.; Vaidyanathan, P.; Paralanov, V.; Strychalski, E.A.; Ross, D.; Densmore, D.; Voigt, C.A. Genetic circuit design automation. *Science* **2016**, *352*, aac7341. [CrossRef] [PubMed]
5. Bhatia, S.P.; Smanski, M.J.; Voigt, C.A.; Densmore, D.M. Genetic Design via Combinatorial Constraint Specification. *ACS Synth. Biol.* **2017**, *6*, 2130–2135. [CrossRef] [PubMed]
6. Garenne, D.; Noireaux, V. Cell-free transcription-translation: Engineering biology from the nanometer to the millimeter scale. *Curr. Opin. Biotechnol.* **2019**, *58*, 19–27. [CrossRef] [PubMed]
7. Dudley, Q.M.; Karim, A.S.; Jewett, M.C. Cell-free metabolic engineering: Biomanufacturing beyond the cell. *Biotechnol. J.* **2015**, *10*, 69–82. [CrossRef]
8. Guo, W.; Sheng, J.; Feng, X. Mini-review: In vitro Metabolic Engineering for Biomanufacturing of High-value Products. *Comput. Struct. Biotechnol. J.* **2017**, *15*, 161–167. [CrossRef]
9. Carlson, E.D.; Gan, R.; Hodgman, C.E.; Jewett, M.C. Cell-free protein synthesis: Applications come of age. *Biotechnol. Adv.* **2012**, *30*, 1185–1194. [CrossRef]
10. Shin, J.; Noireaux, V. An *E. coli* Cell-Free Expression Toolbox: Application to Synthetic Gene Circuits and Artificial Cells. *ACS Synth. Biol.* **2012**, *1*, 29–41. [CrossRef]
11. Noireaux, V.; Bar-Ziv, R.; Libchaber, A. Principles of cell-free genetic circuit assembly. *Proc. Natl. Acad. Sci. USA* **2003**, *100*, 12672–12677. [CrossRef]
12. Marshall, R.; Maxwell, C.S.; Collins, S.P.; Beisel, C.L.; Noireaux, V. Short DNA containing chi sites enhances DNA stability and gene expression in *E. coli* cell-free transcription-translation systems. *Biotechnol. Bioeng.* **2017**, *114*, 2137–2141. [CrossRef] [PubMed]
13. Stiege, W.; Erdmann, V.A. The potentials of the in vitro protein biosynthesis system. *J. Biotechnol.* **1995**, *41*, 81–90. [CrossRef]

14. Jewett, M.C.; Noireaux, V. Synthetic biology: Tailor-made genetic codes. *Nat. Chem.* **2016**, *8*, 291–292. [CrossRef] [PubMed]

15. Timm, A.C.; Shankles, P.G.; Foster, C.M.; Doktycz, M.J.; Retterer, S.T. Toward Microfluidic Reactors for Cell-Free Protein Synthesis at the Point-of-Care. *Small* **2016**, *12*, 810–817. [CrossRef] [PubMed]

16. Noireaux, V.; Maeda, Y.T.; Libchaber, A. Development of an artificial cell, from self-organization to computation and self-reproduction. *Proc. Natl. Acad. Sci. USA* **2011**, *108*, 3473–3480. [CrossRef] [PubMed]

17. Matveev, S.V.; Vinokurov, L.M.; Shaloiko, L.A.; Davies, C.; Matveeva, E.A.; Alakhov, Y.B. Effect of the ATP level on the overall protein biosynthesis rate in a wheat germ cell-free system. *Biochim. Biophys. Acta* **1996**, *1293*, 207–212. [CrossRef]

18. Ge, X.; Luo, D.; Xu, J. Cell-Free Protein Expression under Macromolecular Crowding Conditions. *PLoS ONE* **2011**, *6*, e28707. [CrossRef]

19. Siegal-Gaskins, D.; Tuza, Z.A.; Kim, J.; Noireaux, V.; Murray, R.M. Gene Circuit Performance Characterization and Resource Usage in a Cell-Free "Breadboard". *ACS Synth. Biol.* **2014**, *3*, 416–425. [CrossRef]

20. Shimizu, Y.; Inoue, A.; Tomari, Y.; Suzuki, T.; Yokogawa, T.; Nishikawa, K.; Ueda, T. Cell-free translation reconstituted with purified components. *Nat. Biotechnol.* **2001**, *19*, 751–755. [CrossRef]

21. Pardee, K.; Green, A.A.; Ferrante, T.; Cameron, D.E.; DaleyKeyser, A.; Yin, P.; Collins, J.J. Paper-based Synthetic Gene Networks. *Cell* **2014**, *159*, 940–954. [CrossRef] [PubMed]

22. Garamella, J.; Marshall, R.; Rustad, M.; Noireaux, V. The All *E. coli* TX-TL Toolbox 2.0: A Platform for Cell-Free Synthetic Biology. *ACS Synth. Biol.* **2016**, *5*, 344–355. [CrossRef]

23. Marshall, R.; Maxwell, C.S.; Collins, S.P.; Jacobsen, T.; Luo, M.L.; Begemann, M.B.; Gray, B.N.; January, E.; Singer, A.; He, Y.; et al. Rapid and Scalable Characterization of CRISPR Technologies Using an *E. coli* Cell-Free Transcription-Translation System. *Mol. Cell* **2018**, *69*, 146–157.e3. [CrossRef] [PubMed]

24. Chappell, J.; Jensen, K.; Freemont, P.S. Validation of an entirely in vitro approach for rapid prototyping of DNA regulatory elements for synthetic biology. *Nucleic Acids Res.* **2013**, *41*, 3471–3481. [CrossRef] [PubMed]

25. Sun, Z.Z.; Yeung, E.; Hayes, C.A.; Noireaux, V.; Murray, R.M. Linear DNA for rapid prototyping of synthetic biological circuits in an *Escherichia coli* based TX-TL cell-free system. *ACS Synth. Biol.* **2014**, *3*, 387–397. [CrossRef]

26. Takahashi, M.K.; Chappell, J.; Hayes, C.A.; Sun, Z.Z.; Kim, J.; Singhal, V.; Spring, K.J.; Al-Khabouri, S.; Fall, C.P.; Noireaux, V.; et al. Rapidly characterizing the fast dynamics of RNA genetic circuitry with cell-free transcription-translation (TX-TL) systems. *ACS Synth. Biol.* **2015**, *4*, 503–515. [CrossRef]

27. Niederholtmeyer, H.; Sun, Z.Z.; Hori, Y.; Yeung, E.; Verpoorte, A.; Murray, R.M.; Maerkl, S.J. Rapid cell-free forward engineering of novel genetic ring oscillators. *eLife* **2015**, *4*, e09771. [CrossRef]

28. Lucks, J.B.; Qi, L.; Mutalik, V.K.; Wang, D.; Arkin, A.P. Versatile RNA-sensing transcriptional regulators for engineering genetic networks. *Proc. Natl. Acad. Sci. USA* **2011**, *108*, 8617–8622. [CrossRef]

29. Chappell, J.; Takahashi, M.K.; Meyer, S.; Loughrey, D.; Watters, K.E.; Lucks, J. The centrality of RNA for engineering gene expression. *Biotechnol. J.* **2013**, *8*, 1379–1395. [CrossRef]

30. Rosenfeld, N.; Alon, U. Response Delays and the Structure of Transcription Networks. *J. Mol. Biol.* **2003**, *329*, 645–654. [CrossRef]

31. Shen-Orr, S.S.; Milo, R.; Mangan, S.; Alon, U. Network motifs in the transcriptional regulation network of *Escherichia coli*. *Nat. Genet.* **2002**, *31*, 64–68. [CrossRef] [PubMed]

32. Hu, C.Y.; Takahashi, M.K.; Zhang, Y.; Lucks, J.B. Engineering a Functional Small RNA Negative Autoregulation Network with Model-Guided Design. *ACS Synth. Biol.* **2018**, *7*, 1507–1518. [CrossRef] [PubMed]

33. Shin, J.; Jardine, P.; Noireaux, V. Genome Replication, Synthesis, and Assembly of the Bacteriophage T7 in a Single Cell-Free Reaction. *ACS Synth. Biol.* **2012**, *1*, 408–413. [CrossRef] [PubMed]

34. Rustad, M.; Eastlund, A.; Jardine, P.; Noireaux, V. Cell-free TXTL synthesis of infectious bacteriophage T4 in a single test tube reaction. *Synth. Biol.* **2018**, *3*, ysy002. [CrossRef]

35. Pardee, K.; Green, A.A.; Takahashi, M.K.; Braff, D.; Lambert, G.; Lee, J.W.; Ferrante, T.; Ma, D.; Donghia, N.; Fan, M.; et al. Rapid, Low-Cost Detection of Zika Virus Using Programmable Biomolecular Components. *Cell* **2016**, *165*, 1255–1266. [CrossRef] [PubMed]

36. Pardee, K.; Slomovic, S.; Nguyen, P.Q.; Lee, J.W.; Donghia, N.; Burrill, D.; Ferrante, T.; McSorley, F.R.; Furuta, Y.; Vernet, A.; et al. Portable, On-Demand Biomolecular Manufacturing. *Cell* **2016**, *167*, 248–259.e12. [CrossRef] [PubMed]

37. Schwarz-Schilling, M.; Kim, J.; Cuba, C.; Weitz, M.; Franco, E.; Simmel, F.C. Building a Synthetic Transcriptional Oscillator. *Methods Mol. Biol.* **2016**, *1342*, 185–199. [PubMed]
38. Kim, J.; Hopfield, J.J.; Winfree, E. Neural network computation by in vitro transcriptional circuits. In Proceedings of the 17th International Conference on Neural Information Processing Systems, Vancouver, BC, Canada, 5–8 December 2005; pp. 681–688.
39. Kim, J.; White, K.S.; Winfree, E. Construction of an in vitro bistable circuit from synthetic transcriptional switches. *Mol. Syst. Biol.* **2006**, *2*, 68. [CrossRef]
40. Franco, E.; Friedrichs, E.; Kim, J.; Jungmann, R.; Murray, R.; Winfree, E.; Simmel, F.C. Timing molecular motion and production with a synthetic transcriptional clock. *Proc. Natl. Acad. Sci. USA* **2011**, *108*, E784–E793. [CrossRef]
41. Martin, C.T.; Coleman, J.E. Kinetic Analysis of T7 RNA Polymerase-Promoter Interactions with Small Synthetic Promoters. *Biochemistry* **1987**, *26*, 2690–2696. [CrossRef]
42. Subsoontorn, P.; Kim, J.; Winfree, E. Ensemble Bayesian Analysis of Bistability in a Synthetic Transcriptional Switch. *ACS Synth. Biol.* **2012**, *1*, 299–316. [CrossRef] [PubMed]
43. Yurke, B.; Mills, A.P., Jr. Using DNA to power nanostructures. *Genet. Program. Evolvable Mach.* **2003**, *4*, 111–122. [CrossRef]
44. Markevich, N.I.; Hoek, J.B.; Kholodenko, B.N. Signaling switches and bistability arising from multisite phosphorylation in protein kinase cascades. *J. Cell Biol.* **2004**, *164*, 353–359. [CrossRef] [PubMed]
45. Ghaffarizadeh, A.; Flann, N.S.; Podgorski, G.J. Multistable switches and their role in cellular differentiation networks. *BMC Bioinform.* **2014**, *15*, S7. [CrossRef] [PubMed]
46. Proctor, C.J.; Gray, D.A. Explaining oscillations and variability in the p53-Mdm2 system. *BMC Syst. Biol.* **2008**, *2*, 75. [CrossRef] [PubMed]
47. Nelson, D.E.; Ihekwaba, A.E.C.; Elliott, M.; Johnson, J.R.; Gibney, C.A.; Foreman, B.E.; Nelson, G.; See, V.; Horton, C.A.; Spiller, D.G.; et al. Oscillations in NF-κB Signaling Control the Dynamics of Gene Expression. *Science* **2004**, *306*, 704–708. [CrossRef]
48. Kim, J.; Winfree, E. Synthetic in vitro transcriptional oscillators. *Mol. Syst. Biol.* **2011**, *7*, 465. [CrossRef]
49. Cho, E.J.; Lee, J.-W.; Ellington, A.D. Applications of Aptamers as Sensors. *Annu. Rev. Anal. Chem.* **2009**, *2*, 241–264. [CrossRef]
50. Dupin, A.; Simmel, F.C. Signalling and differentiation in emulsion-based multi-compartmentalized in vitro gene circuits. *Nat. Chem.* **2019**, *11*, 32–39. [CrossRef]
51. Lloyd, J.; Tran, C.H.; Wadhwani, K.; Cuba Samaniego, C.; Subramanian, H.K.K.; Franco, E. Dynamic Control of Aptamer-Ligand Activity Using Strand Displacement Reactions. *ACS Synth. Biol.* **2018**, *7*, 30–37. [CrossRef]
52. Kim, J.; Quijano, J.F.; Yeung, E.; Murray, R.M. Synthetic logic circuits using RNA aptamer against T7 RNA polymerase. *bioRxiv* **2014**. [CrossRef]
53. Cuba Samaniego, C.; Franco, E. A Robust Molecular Network Motif for Period-Doubling Devices. *ACS Synth. Biol.* **2018**, *7*, 75–85. [CrossRef] [PubMed]
54. Franco, E.; Giordano, G.; Forsberg, P.-O.; Murray, R.M. Negative Autoregulation Matches Production and Demand in Synthetic Transcriptional Networks. *ACS Synth. Biol.* **2014**, *3*, 589–599. [CrossRef] [PubMed]
55. Kim, J.; Khetarpal, I.; Sen, S.; Murray, R.M. Synthetic circuit for exact adaptation and fold-change detection. *Nucleic Acids Res.* **2014**, *42*, 6078–6089. [CrossRef] [PubMed]
56. Cuba Samaniego, C.; Giordano, G.; Kim, J.; Blanchini, F.; Franco, E. Molecular Titration Promotes Oscillations and Bistability in Minimal Network Models with Monomeric Regulators. *ACS Synth. Biol.* **2016**, *5*, 321–333. [CrossRef] [PubMed]
57. Ayukawa, S.; Takinoue, M.; Kiga, D. RTRACS: A Modularized RNA-Dependent RNA Transcription System with High Programmability. *Acc. Chem. Res.* **2011**, *44*, 1369–1379. [CrossRef] [PubMed]
58. Takinoue, M.; Kiga, D.; Shohda, K.-I.; Suyama, A. Experiments and simulation models of a basic computation element of an autonomous molecular computing system. *Phys. Rev. E Stat. Nonlinear Soft Matter Phys.* **2008**, *78*, 041921. [CrossRef]
59. Montagne, K.; Plasson, R.; Sakai, Y.; Fujii, T.; Rondelez, Y. Programming an in vitro DNA oscillator using a molecular networking strategy. *Mol. Syst. Biol.* **2011**, *7*, 466. [CrossRef]
60. Baccouche, A.; Montagne, K.; Padirac, A.; Fujii, T.; Rondelez, Y. Dynamic DNA-toolbox reaction circuits: A walkthrough. *Methods* **2014**, *67*, 234–249. [CrossRef]

61. Padirac, A.; Fujii, T.; Rondelez, Y. Bottom-up construction of in vitro switchable memories. *Proc. Natl. Acad. Sci. USA* **2012**, *109*, E3212–E3220. [CrossRef]

62. Fujii, T.; Rondelez, Y. Predator-prey molecular ecosystems. *ACS Nano* **2013**, *7*, 27–34. [CrossRef] [PubMed]

63. Zadorin, A.S.; Rondelez, Y.; Gines, G.; Dilhas, V.; Urtel, G.; Zambrano, A.; Galas, J.-C.; Estevez-Torres, A. Synthesis and materialization of a reaction–diffusion French flag pattern. *Nat. Chem.* **2017**, *9*, 990–996. [CrossRef] [PubMed]

64. Padirac, A.; Fujii, T.; Rondelez, Y. Quencher-free multiplexed monitoring of DNA reaction circuits. *Nucleic Acids Res.* **2012**, *40*, e118. [CrossRef] [PubMed]

65. Soloveichik, D.; Seelig, G.; Winfree, E. DNA as a universal substrate for chemical kinetics. *Proc. Natl. Acad. Sci. USA* **2010**, *107*, 5393–5398. [CrossRef] [PubMed]

66. Seelig, G.; Soloveichik, D.; Zhang, D.Y.; Winfree, E. Enzyme-Free Nucleic Acid Logic Circuits. *Science* **2006**, *314*, 1585–1588. [CrossRef]

67. Qian, L.; Winfree, E. Scaling Up Digital Circuit Computation with DNA Strand Displacement Cascades. *Science* **2011**, *332*, 1196–1201. [CrossRef] [PubMed]

68. Qian, L.; Winfree, E.; Bruck, J. Neural network computation with DNA strand displacement cascades. *Nature* **2011**, *475*, 368–372. [CrossRef]

69. Cherry, K.M.; Qian, L. Scaling up molecular pattern recognition with DNA-based winner-take-all neural networks. *Nature* **2018**, *559*, 370–376. [CrossRef]

70. Srinivas, N.; Parkin, J.; Seelig, G.; Winfree, E.; Soloveichik, D. Enzyme-free nucleic acid dynamical systems. *Science* **2017**, *358*, eaal2052. [CrossRef]

71. Douglas, S.M.; Bachelet, I.; Church, G.M. A Logic-Gated Nanorobot for Targeted Transport of Molecular Payloads. *Science* **2012**, *335*, 831–834. [CrossRef]

72. Isaacs, F.J.; Dwyer, D.J.; Collins, J.J. RNA synthetic biology. *Nat Biotechnol.* **2006**, *24*, 545–554. [CrossRef] [PubMed]

73. Lease, R.A.; Belfort, M. A trans-acting RNA as a control switch in *Escherichia coli*: DsrA modulates function by forming alternative structures. *Proc. Natl. Acad. Sci. USA* **2000**, *97*, 9919–9924. [CrossRef] [PubMed]

74. Isaacs, F.J.; Dwyer, D.J.; Ding, C.; Pervouchine, D.D.; Cantor, C.R.; Collins, J.J. Engineered riboregulators enable post-transcriptional control of gene expression. *Nat. Biotechnol.* **2004**, *22*, 841–847. [CrossRef] [PubMed]

75. Green, A.A.; Silver, P.A.; Collins, J.J.; Yin, P. Toehold Switches: De-Novo-Designed Regulators of Gene Expression. *Cell* **2014**, *159*, 925–939. [CrossRef] [PubMed]

76. Ma, D.; Shen, L.; Wu, K.; Diehnelt, C.W.; Green, A.A. Low-cost detection of norovirus using paper-based cell-free systems and synbody-based viral enrichment. *Synth. Biol.* **2018**, *3*, ysy018. [CrossRef] [PubMed]

77. Takahashi, M.K.; Tan, X.; Dy, A.J.; Braff, D.; Akana, R.T.; Furuta, Y.; Donghia, N.; Ananthakrishnan, A.; Collins, J.J. A low-cost paper-based synthetic biology platform for analyzing gut microbiota and host biomarkers. *Nat. Commun.* **2018**, *9*, 3347. [CrossRef] [PubMed]

78. Kim, J.; Yin, P.; Green, A.A. Ribocomputing: Cellular Logic Computation Using RNA Devices. *Biochemistry* **2018**, *57*, 883–885. [CrossRef] [PubMed]

79. Green, A.A.; Kim, J.; Ma, D.; Silver, P.A.; Collins, J.J.; Yin, P. Complex cellular logic computation using ribocomputing devices. *Nature* **2017**, *548*, 117–121. [CrossRef]

80. Kim, J.; Zhou, Y.; Carlson, P.; Teichmann, M.; Simmel, F.C.; Silver, P.A.; Collins, J.J.; Lucks, J.B.; Yin, P.; Green, A.A. De-Novo-Designed Translational Repressors for Multi-Input Cellular Logic. *bioRxiv* **2018**. [CrossRef]

81. Carlson, P.D.; Glasscock, C.J.; Lucks, J.B. De novo Design of Translational RNA Repressors. *bioRxiv* **2018**. [CrossRef]

82. Takahashi, M.K.; Lucks, J.B. A modular strategy for engineering orthogonal chimeric RNA transcription regulators. *Nucleic Acids Res.* **2013**, *41*, 7577–7588. [CrossRef] [PubMed]

83. Chappell, J.; Takahashi, M.K.; Lucks, J.B. Creating small transcription activating RNAs. *Nat. Chem. Biol.* **2015**, *11*, 214–220. [CrossRef] [PubMed]

84. Chappell, J.; Westbrook, A.; Verosloff, M.; Lucks, J.B. Computational design of small transcription activating RNAs for versatile and dynamic gene regulation. *Nat. Commun.* **2017**, *8*, 1051. [CrossRef]

85. Verosloff, M.; Chappell, J.; Perry, K.L.; Thompson, J.R.; Lucks, J.B. PLANT-Dx: A Molecular Diagnostic for Point of Use Detection of Plant Pathogens. *ACS Synth. Biol.* **2019**, *8*, 902–905. [CrossRef] [PubMed]

86. Hoynes-O'Connor, A.; Hinman, K.; Kirchner, L.; Moon, T.S. De novo design of heat-repressible RNA thermosensors in *E. coli*. *Nucleic Acids Res.* **2015**, *43*, 6166–6179. [CrossRef]

87. Lee, Y.J.; Moon, T.S. Design rules of synthetic non-coding RNAs in bacteria. *Methods* **2018**, *143*, 58–69. [CrossRef] [PubMed]

88. Rodrigo, G.; Jaramillo, A. AutoBioCAD: Full biodesign automation of genetic circuits. *ACS Synth. Biol.* **2013**, *2*, 230–236. [CrossRef]

89. Rodrigo, G.; Landrain, T.E.; Jaramillo, A. De novo automated design of small RNA circuits for engineering synthetic riboregulation in living cells. *Proc. Natl. Acad. Sci. USA* **2012**, *109*, 15271–15276. [CrossRef]

90. Noireaux, V.; Libchaber, A. A vesicle bioreactor as a step toward an artificial cell assembly. *Proc. Natl. Acad. Sci. USA* **2004**, *101*, 17669–17674. [CrossRef]

91. Minton, A.P. The Influence of Macromolecular Crowding and Macromolecular Confinement on Biochemical Reactions in Physiological Media. *J. Biol. Chem.* **2001**, *276*, 10577–10580. [CrossRef]

92. Holtze, C.; Rowat, A.C.; Agresti, J.J.; Hutchison, J.B.; Angilè, F.E.; Schmitz, C.H.J.; Köster, S.; Duan, H.; Humphry, K.J.; Scanga, R.A.; et al. Biocompatible surfactants for water-in-fluorocarbon emulsions. *Lab Chip* **2008**, *8*, 1632. [CrossRef] [PubMed]

93. Walter, A.; Gutknecht, J. Permeability of small nonelectrolytes through lipid bilayer membranes. *J. Membr. Biol.* **1986**, *90*, 207–217. [CrossRef] [PubMed]

94. Finkelstein, A. Water and nonelectrolyte permeability of lipid bilayer membranes. *J. Gen. Physiol.* **1976**, *68*, 127–135. [CrossRef] [PubMed]

95. Wei, C.; Pohorille, A. Permeation of Membranes by Ribose and Its Diastereomers. *J. Am. Chem. Soc.* **2009**, *131*, 10237–10245. [CrossRef] [PubMed]

96. Abkarian, M.; Loiseau, E.; Massiera, G. Continuous droplet interface crossing encapsulation (cDICE) for high throughput monodisperse vesicle design. *Soft Matter* **2011**, *7*, 4610. [CrossRef]

97. Ishikawa, K.; Sato, K.; Shima, Y.; Urabe, I.; Yomo, T. Expression of a cascading genetic network within liposomes. *FEBS Lett.* **2004**, *576*, 387–390. [CrossRef]

98. Ota, S.; Yoshizawa, S.; Takeuchi, S. Microfluidic Formation of Monodisperse, Cell-Sized, and Unilamellar Vesicles. *Angew. Chem.* **2009**, *121*, 6655–6659. [CrossRef]

99. Weiss, M.; Frohnmayer, J.P.; Benk, L.T.; Haller, B.; Janiesch, J.W.; Heitkamp, T.; Börsch, M.; Lira, R.B.; Dimova, R.; Lipowsky, R.; et al. Sequential bottom-up assembly of mechanically stabilized synthetic cells by microfluidics. *Nat. Mater.* **2018**, *17*, 89–96. [CrossRef]

100. Matosevic, S.; Paegel, B.M. Stepwise Synthesis of Giant Unilamellar Vesicles on a Microfluidic Assembly Line. *J. Am. Chem. Soc.* **2011**, *133*, 2798–2800. [CrossRef]

101. Tan, C.; Saurabh, S.; Bruchez, M.P.; Schwartz, R.; LeDuc, P. Molecular crowding shapes gene expression in synthetic cellular nanosystems. *Nat. Nanotechnol.* **2013**, *8*, 602–608. [CrossRef]

102. Hansen, M.M.; Meijer, L.H.; Spruijt, E.; Maas, R.J.; Rosquelles, M.V.; Groen, J.; Heus, H.A.; Huck, W.T.S. Macromolecular crowding creates heterogeneous environments of gene expression in picolitre droplets. *Nat. Nanotechnol.* **2016**, *11*, 191–197. [CrossRef] [PubMed]

103. Torre, P.; Keating, C.D.; Mansy, S.S. Multiphase Water-in-Oil Emulsion Droplets for Cell-Free Transcription–Translation. *Langmuir* **2014**, *30*, 5695–5699. [CrossRef]

104. Weitz, M.; Kim, J.; Kapsner, K.; Winfree, E.; Franco, E.; Simmel, F.C. Diversity in the dynamical behaviour of a compartmentalized programmable biochemical oscillator. *Nat. Chem.* **2014**, *6*, 295–302. [CrossRef] [PubMed]

105. Adamala, K.P.; Martin-Alarcon, D.A.; Guthrie-Honea, K.R.; Boyden, E.S. Engineering genetic circuit interactions within and between synthetic minimal cells. *Nat. Chem.* **2017**, *9*, 431–439. [CrossRef] [PubMed]

106. de Souza, T.P.; Fahr, A.; Luisi, P.L.; Stano, P. Spontaneous encapsulation and concentration of biological macromolecules in liposomes: An intriguing phenomenon and its relevance in origins of life. *J. Mol. Evol.* **2014**, *79*, 179–192. [CrossRef] [PubMed]

107. Klumpp, S.; Scott, M.; Pedersen, S.; Hwa, T. Molecular crowding limits translation and cell growth. *Proc. Natl. Acad. Sci. USA* **2013**, *110*, 16754–16759. [CrossRef]

108. Albertsson, P.A. Partition of cell particles and macromolecules in polymer two-phase systems. *Adv. Protein Chem.* **1970**, *24*, 309–341.

109. Aumiller, W.M., Jr.; Keating, C.D. Experimental models for dynamic compartmentalization of biomolecules in liquid organelles: Reversible formation and partitioning in aqueous biphasic systems. *Adv. Colloid Interface Sci.* **2017**, *239*, 75–87. [CrossRef]

110. Hyman, A.A.; Weber, C.A.; Jülicher, F. Liquid-Liquid Phase Separation in Biology. *Annu. Rev. Cell Dev. Biol.* **2014**, *30*, 39–58. [CrossRef]

111. Lentini, R.; Martín, N.Y.; Forlin, M.; Belmonte, L.; Fontana, J.; Cornella, M.; Martini, L.; Tamburini, S.; Bentley, W.E.; Jousson, O.; et al. Two-Way Chemical Communication between Artificial and Natural Cells. *ACS Cent. Sci.* **2017**, *3*, 117–123. [CrossRef]

112. Bayoumi, M.; Bayley, H.; Maglia, G.; Sapra, K.T. Multi-compartment encapsulation of communicating droplets and droplet networks in hydrogel as a model for artificial cells. *Sci. Rep.* **2017**, *7*, 45167. [CrossRef] [PubMed]

113. Elani, Y.; Trantidou, T.; Wylie, D.; Dekker, L.; Polizzi, K.; Law, R.V.; Ces, O. Constructing vesicle-based artificial cells with embedded living cells as organelle-like modules. *Sci. Rep.* **2018**, *8*, 4564. [CrossRef] [PubMed]

114. Rampioni, G.; Leoni, L.; Mavelli, F.; Damiano, L.; Stano, P. Interfacing Synthetic Cells with Biological Cells: An Application of the Synthetic Method. In Proceedings of the 2018 Conference on Artificial Life: A Hybrid of the European Conference on Artificial Life (ECAL) and the International Conference on the Synthesis and Simulation of Living Systems (ALIFE), Tokyo, Japan, 23–27 July 2018; pp. 145–146.

115. Lentini, R.; Santero, S.P.; Chizzolini, F.; Cecchi, D.; Fontana, J.; Marchioretto, M.; Del Bianco, C.; Terrell, J.L.; Spencer, A.C.; Martini, L.; et al. Integrating artificial with natural cells to translate chemical messages that direct *E. coli* behaviour. *Nat. Commun.* **2014**, *5*, 4012. [CrossRef] [PubMed]

116. Bloomfield, V.A.; Crothers, D.M.; Tinoco, I.; Hearst, J.E.; Wemmer, D.E.; Killman, P.A.; Turner, D.H. *Nucleic Acids: Structures, Properties, and Functions*, 1st ed.; University Science Books: Sausalito, CA, USA, 2000; 627p.

117. SantaLucia, J.; Hicks, D. The Thermodynamics of DNA Structural Motifs. *Annu. Rev. Biophys. Biomol. Struct.* **2004**, *33*, 415–440. [CrossRef] [PubMed]

118. Rothemund, P.W.K.; Ekani-Nkodo, A.; Papadakis, N.; Kumar, A.; Fygenson, D.K.; Winfree, E. Design and Characterization of Programmable DNA Nanotubes. *J. Am. Chem. Soc.* **2004**, *126*, 16344–16352. [CrossRef] [PubMed]

119. Li, W.; Bell, N.A.W.; Hernández-Ainsa, S.; Thacker, V.V.; Thackray, A.M.; Bujdoso, R.; Keyser, U.F. Single Protein Molecule Detection by Glass Nanopores. *ACS Nano* **2013**, *7*, 4129–4134. [CrossRef] [PubMed]

120. Zhang, D.Y.; Seelig, G. Dynamic DNA nanotechnology using strand-displacement reactions. *Nat. Chem.* **2011**, *3*, 103–113. [CrossRef] [PubMed]

121. Yurke, B.; Turberfield, A.J.; Mills, A.P., Jr.; Simmel, F.C.; Neumann, J.L. A DNA-fuelled molecular machine made of DNA. *Nature* **2000**, *406*, 605–608. [CrossRef]

122. Dirks, R.M.; Pierce, N.A. Triggered amplification by hybridization chain reaction. *Proc. Natl. Acad. Sci. USA* **2004**, *101*, 15275–15278. [CrossRef]

123. Keum, J.W.; Bermudez, H. Enhanced resistance of DNA nanostructures to enzymatic digestion. *Chem. Commun.* **2009**, 7036–7038. [CrossRef]

124. Castro, C.E.; Kilchherr, F.; Kim, D.N.; Shiao, E.L.; Wauer, T.; Wortmann, P.; Bathe, M.; Dietz, H. A primer to scaffolded DNA origami. *Nat. Methods* **2011**, *8*, 221–229. [CrossRef] [PubMed]

125. Mei, Q.; Wei, X.; Su, F.; Liu, Y.; Youngbull, C.; Johnson, R.; Lindsay, S.; Yan, H.; Meldrum, D. Stability of DNA Origami Nanoarrays in Cell Lysate. *Nano Lett.* **2011**, *11*, 1477–1482. [CrossRef] [PubMed]

126. Klocke, M.A.; Garamella, J.; Subramanian, H.K.K.; Noireaux, V.; Franco, E. Engineering DNA nanotubes for resilience in an *E. coli* TXTL system. *Synth. Biol.* **2018**, *3*, ysy001. [CrossRef]

127. Bramsen, J.B.; Laursen, M.B.; Nielsen, A.F.; Hansen, T.B.; Bus, C.; Langkjær, N.; Babu, B.R.; Højland, T.; Abramov, M.; Van Aerschot, A.; et al. A large-scale chemical modification screen identifies design rules to generate siRNAs with high activity, high stability and low toxicity. *Nucleic Acids Res.* **2009**, *37*, 2867–2881. [CrossRef] [PubMed]

128. Conway, J.W.; McLaughlin, C.K.; Castor, K.J.; Sleiman, H. DNA nanostructure serum stability: Greater than the sum of its parts. *Chem. Commun.* **2013**, *49*, 1172. [CrossRef]

129. Hahn, J.; Wickham, S.F.J.; Shih, W.M.; Perrault, S.D. Addressing the Instability of DNA Nanostructures in Tissue Culture. *ACS Nano* **2014**, *8*, 8765–8775. [CrossRef] [PubMed]

130. Goltry, S.; Hallstrom, N.; Clark, T.; Kuang, W.; Lee, J.; Jorcyk, C.; Knowlton, W.B.; Yurke, B.; Hughes, W.L.; Graugnard, E. DNA topology influences molecular machine lifetime in human serum. *Nanoscale* **2015**, *7*, 10382–10390. [CrossRef]

131. Zhang, D.Y.; Turberfield, A.J.; Yurke, B.; Winfree, E. Engineering Entropy-Driven Reactions and Networks Catalyzed by DNA. *Science* **2007**, *318*, 1121–1125. [CrossRef]
132. Graugnard, E.; Cox, A.; Lee, J.; Jorcyk, C.; Yurke, B.; Hughes, W.L. Operation of a DNA-Based Autocatalytic Network in Serum. In Proceedings of the 16th International Conference on DNA Computing and Molecular Programming, Hong Kong, China, 14–17 June 2010; pp. 83–88.
133. Fern, J.; Schulman, R. Design and Characterization of DNA Strand-Displacement Circuits in Serum-Supplemented Cell Medium. *ACS Synth. Biol.* **2017**, *6*, 1774–1783. [CrossRef]
134. Schaffter, S.W.; Green, L.N.; Schneider, J.; Subramanian, H.K.K.; Schulman, R.; Franco, E. T7 RNA polymerase non-specifically transcribes and induces disassembly of DNA nanostructures. *Nucleic Acids Res.* **2018**, *46*, 5332–5343. [CrossRef]
135. Montagne, K.; Gines, G.; Fujii, T.; Rondelez, Y. Boosting functionality of synthetic DNA circuits with tailored deactivation. *Nat. Commun.* **2016**, *7*, 13474. [CrossRef] [PubMed]
136. Westbrook, A.; Tang, X.; Marshall, R.; Maxwell, C.S.; Chappell, J.; Agrawal, D.K.; Dunlop, M.J.; Noireaux, V.; Beisel, C.L.; Lucks, J.; et al. Distinct timescales of RNA regulators enable the construction of a genetic pulse generator. *Biotechnol. Bioeng.* **2019**, *116*, 1139–1151. [CrossRef] [PubMed]
137. Hu, C.Y.; Varner, J.D.; Lucks, J.B. Generating Effective Models and Parameters for RNA Genetic Circuits. *ACS Synth. Biol.* **2015**, *4*, 914–926. [CrossRef] [PubMed]
138. Joesaar, A.; Yang, S.; Bogels, B.; van der Linden, A.; Pieters, P.; Kumar, B.V.V.S.P.; Dalchau, N.; Phillips, A.; Mann, S.; de Greef, T.F.A. Distributed DNA-based Communication in Populations of Synthetic Protocells. *Nat. Nanotechnol.* **2019**, *14*, 369–378. [CrossRef] [PubMed]

methods
and
protocols

MDPI

Protocol

Efficient Incorporation of Unnatural Amino Acids into Proteins with a Robust Cell-Free System

Wei Gao [1,2] (ID), **Ning Bu** [1,*] **and Yuan Lu** [2,3,4,*] (ID)

1   College of Life Science, Shenyang Normal University, Shenyang 100034, Liaoning, China;
    weigao1122@gmail.com
2   Department of Chemical Engineering, Tsinghua University, Beijing 100084, China
3   Institute of Biochemical Engineering, Department of Chemical Engineering, Tsinghua University,
    Beijing 100084, China
4   Key Lab of Industrial Biocatalysis, Ministry of Education, Department of Chemical Engineering, Tsinghua
    University, Beijing 100084, China
*   Correspondence: buning@synu.edu.cn (N.B.); yuanlu@tsinghua.edu.cn (Y.L.); Tel.: +86-10-62780127 (Y.L.)

Received: 16 January 2019; Accepted: 7 February 2019; Published: 12 February 2019

Abstract: Unnatural proteins are crucial biomacromolecules and have been widely applied in fundamental science, novel biopolymer materials, enzymes, and therapeutics. Cell-free protein synthesis (CFPS) system can serve as a robust platform to synthesize unnatural proteins by highly effective site-specific incorporation of unnatural amino acids (UNAAs), without the limitations of cell membrane permeability and the toxicity of unnatural components. Here, we describe a quick and simple method to synthesize unnatural proteins in CFPS system based on *Escherichia coli* crude extract, with unnatural orthogonal aminoacyl-tRNA synthetase and suppressor tRNA evolved from *Methanocaldococcus jannaschii*. The superfolder green fluorescent protein (sfGFP) and *p*-propargyloxyphenylalanine (*p*PaF) were used as the model protein and UNAA. The synthesis of unnatural sfGFPs was characterized by microplate spectrophotometer, affinity chromatography, and liquid chromatography-mass spectrometry/mass spectrometry (LC-MS/MS). This protocol provides a detailed procedure guiding how to use the powerful CFPS system to synthesize unnatural proteins on demand.

Keywords: cell-free protein synthesis; unnatural amino acid; unnatural protein

1. Introduction

Protein is a vital class of biomolecules, and all living organisms employ it to fulfill essential structural, functional, and enzymatic roles to sustain life [1]. In nature, 20 natural amino acids can form proteins in a near-infinite number of combinations to make them have structural and functional diversity [1]. However, these natural amino acids can possess some interesting chemistries [2,3]. Thus, expanding protein functions by incorporating unnatural amino acids (UNAAs) featuring novel functional groups is more important [4].

Methods of UNAA incorporation usually comprise global suppression, amber suppression, frame-shift suppression, sense codon reassignment, and unnatural base-pairs; the most popular approach is amber suppression [5]. Usually, unnatural orthogonal translation systems (OTS) are designed to incorporate UNAAs in vivo or in vitro [6,7]. OTS, is an orthogonal system which can covalently load UNAA onto the suppressor tRNA (o-tRNA) via acylation of aminoacyl-tRNA synthetase (o-aaRS) [1]. To date, many various UNAAs have been incorporated into proteins successfully by amber suppression with OTSs [1,8]. Thus, expanding the genetic code by incorporating UNAAs has become a significant opportunity in synthetic and chemical biology [3]. In addition, unnatural proteins have been applied in protein modifications [9], biophysical probes [10], enzyme engineering [11], biomaterials [12] and biopharmaceutical protein production [13]. The synthesis of unnatural proteins can be based on two systems. One is an in vivo cellular system, and the other is an in vitro cell-free protein synthesis (CFPS) system. In conventional in vivo systems, incorporation of UNAAs is dependent on cell viability, and the cell membrane barrier limits the transportation of UNAAs [14]. Compared with in vivo system, a CFPS system has several apparent advantages [7]. A CFPS system consists of crude extract with basal transcription and translation elements, amino acids, DNA templates, energy regeneration substrates, nucleotides, salts, and cofactors [15]. The transcription and translation processes happen in an open biological reaction system. Because there is no limitation of the cell membrane and no cell growth, the CFPS system has higher utilization efficiency of UNAAs, higher flexibility of reactions, higher toxicity tolerance, shorter production cycle [16], and higher protein yield [17]. Therefore, we can combine the CFPS system with the OTS system to highly effective produce unnatural proteins with UNAA by amber suppression.

2. Experimental Design

This protocol describes how to synthesize unnatural proteins with UNAAs. This method is based on a robust CFPS platform, and combines the expansion of the genetic code with unnatural OTSs to achieve the incorporation of UNAAs. In this study, the superfolder green fluorescent protein (sfGFP) was used as the model protein, in which position 2 was chosen for the UNAA incorporation. In this protocol, the incorporation of UNAAs can be preliminarily determined via the fluorescence intensity of unnatural sfGFP; thus, the selected position must not destroy the chromophore. Additionally, a previous study indicated that the yield of unnatural protein that contains UNAA at position 2 is approximate to the yield of the wild-type protein [11]. The experiment is carried out in four steps (See Figure 1). First, crude cell lysate was made. Second, OTS including aminoacyl-tRNA synthetase of *p*-propargyloxyphenylalanine (*p*PaFRS) and o-tDNA^{opt} was made. Third, CFPS reaction components were formulated. Lastly, unnatural proteins were synthesized and characterized. The following section describes the specific experimental procedures.

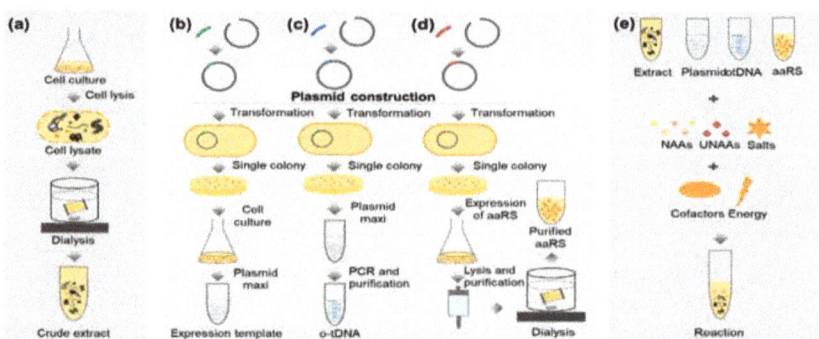

Figure 1. Experimental design. (**a**) Preparation of crude extract; (**b**) extraction of expression templates; (**c**) purification of o-tDNA; (**d**) purification of aaRS; (**e**) cell-free protein synthesis reaction.

2.1. Materials

2.1.1. Preparation of *E. coli* Extract

- *E. coli* Rosetta (DE3) strain (Biomed, Beijing, China; Cat. no.: BC204-01)
- Antifoam 204 (Sigma-Aldrich, St. Louis, MO, USA; Cat. no.: SLBW1473)
- Chloramphenicol (Genview, Beijing, China; Cat. no.: AC060)
- Tryptone (OXIOD, Basingstoke, UK; Cat. no.: LP0042)
- Yeast extract (OXIOD; Cat. no.: LP0021)

- Bacto^{TM} agar (Becton. Dickinson, Franklin Lakes, NJ, USA; Cat. no.: 7291815)
- NaCl (Sinopharm Chemical Reagent, Shanghai, China; Cat. no.: 10019318)
- Dipotassium hydrogen phosphate (KH$_2$PO$_4$) (Tong Guang Fine Chemicals, Beijing, China; Cat. no.: 7778-77-0)
- Potassium hydrogen phosphate trihydrate (K$_2$HPO$_4$·3H$_2$O) (Tong Guang Fine Chemicals; Cat. no.: 16788-57-1)
- Ampicillin (Solarbio, Beijing, China; Cat. no.: A8180)
- Potassium L-glutamate (Yuanye, Shanghai, China; Cat. no.: S20427)
- L-Glutamic acid hemimagnesium salt tetrahydrate (Sigma-Aldrich; Cat. no.: 49605)
- Tris (Biotopped, Beijing, China; Cat. no.: T6061)
- 1,4-Dithio-DL-threitol (DTT) (Solarbio; Cat. no.: D8220)

2.1.2. Preparation of *p*PaFRS

- The nucleotide sequence of *p*PaFRS refers to pPRMjRS-1 [18]
- Plasmid pEVOL-pAzF [19] (pEVOL-pAzF was a gift from Peter Schultz (Addgene plasmid #31186; http://n2t.net/addgene:31186; RRID: Addgene_31186))
- Plasmid pET24a (Novagen, Shanghai, China; Cat. no.: 69749-3)
- Primers P1f, P1r, P2f, P2r, P3f and P3r (See Table S1)
- Q5® High-Fidelity DNA Polymerases (New England Biolabs, Beijing, China; Cat. no.: M0491)
- *E. coli* strain expressing the *p*PaFRS gene: BL21(DE3) competent cells (Biomed; Cat. no.: BC201)
- Kanamycin (Solarbio; Cat. no.: K8020)
- Tryptone (OXIOD; Cat. no.: LP0042)
- Yeast extract (OXIOD; Cat. no.: LP0021)
- BactoTM agar (Becton. Dickinson; Cat. no.: 7291815)
- NaCl (Sinopharm Chemical Reagent; Cat. no.: 10019318)
- Isopropyl-b-D-thiogalactoside (Solarbio; Cat. no.: I8070)
- EzFast Ni HP) columns (5 mL) (BestChrom, Shanghai, China; Cat. no.: EA005)
- Ethanol (TONG GUANG FINE CHEMICALS; Cat. no.: 32061)
- Imidazole (Sigma-Aldrich; Cat. no.: I2399)
- Quick Start Bradford Protein Assay Kit (Bio-Rad, Hercules, CA, USA; Cat. No.: 5000201)
- PBS buffer (Solarbio; Cat. no.: P1010)

2.1.3. Preparation of Crude T7 RNA Polymerase

- Plasmid pAR1219 (Sigma-Aldrich; Cat. no.: T2076).
- *E. coli* strain expressing the T7 RNA polymerase gene: BL21(DE3) competent cells (Biomed; Cat. no.: BC201)
- Tryptone (OXIOD; Cat. no.: LP0042)
- Yeast extract (OXIOD; Cat. no.: LP0021)
- BactoTM agar (Becton. Dickinson; Cat. no.: 7291815)
- NaCl (Sinopharm Chemical Reagent; Cat. no.: 10019318)
- Potassium acetate (Sigma-Aldrich; Cat. no.: V900213)
- Magnesium acetate tetrahydrate (Sigma-Aldrich; Cat. no.: V900172)
- Ethylenediaminetetraacetic acid (EDTA) (Solarbio; Cat. no.: E8040)
- 1,4-Dithio-DL-threitol (DTT) (Solarbio; Cat. no.: D8220)
- Potassium hydrogen phosphate trihydrate (K$_2$HPO$_4$·3H$_2$O) (Tong Guang Fine Chemicals; Cat. no.: 16788-57-1)
- β-mercaptoethanol (Amresco, Shanghai, China; Cat. no.: 0482)

- 1 × Protease inhibitor (Sigma-Aldrich; Cat. no.: P8340)

2.1.4. Preparation of Expression Template and o-tDNA^{opt}

- Plasmid pET23a (Novagen; Cat. no.: 69745-3)
- Primers P4f and P4r (See Table S1) to generate pET23a-sfGFP-StrepII gene.
- Primers P5f and P5r (See Table S1) to generate pET23a-sfGFP(2TAG)-StrepII gene which contained the TAG site in the 2nd codon and C-terminal StrepII tag.
- QIAGEN Plasmid Maxi Kit (10) (QIAGEN, Shanghai, China; Cat. no.: 12162)
- Plasmid Mini Kit (OMEGA Bio-Tek, Atlanta, GA, USA)
- The o-tDNA^{opt} gene (GENEWIZ, Suzhou, China)
- Primers P6f and P6r to amplify the pET23a vector gene (See Table S1)
- Primers P7f and P7r to amplify the single o-tDNA^{opt} gene [20] which can ligate with pET23a vector gene (See Table S1)
- Pfu polymerase (Beyotime Biotechnology, Shanghai, China; Cat. no.: D7217)
- Ethanol (Tong Guang Fine Chemicals; Cat. no.: 32061)
- 3 M Sodium acetate (Solarbio; Cat. no.: A1070)

2.1.5. Synthesis and Characterization of sfGFP and sfGFP2*p*PaF

All the chemicals were from Sigma except special description.

- 10 × salt: 1.75 M Potassium glutamate, 27 mM Potassium oxalate monohydrate, 100 mM Glutamate, adjusted the pH to 7.2~7.4 with ammonia while dissolving
- Mg^{2+} solution: 1 M Magnesium glutamate.
- Amino acids mixture: Added the amino acids in the following order (given in three-letter code): Arg, Val, Trp, Phe, Ile, Leu, Cys, Met, Ala, Asn, Asp, Gly, Gln, His, Lys, Pro, Ser, Thr, Tyr. During preparation, it was necessary to ensure that any of amino acids has been dissolved before adding the next, and Tyr was added at last. Adjusted the pH to 7.4 with ammonia hydroxide.
- ⚠ **CRITICAL STEP** Phosphoenolpyruvic acid (PEP) solution: (Prepare rapidly on the ice and flash frozen) 883 mM Phosphoenolpyruvate, added 10 M KOH to adjust pH to 7.4.
- GSSG (Oxidized glutathione) and GSH (Reduced glutathione): 100 mM GSSG, 100 mM GSH.
- 25 × NTPs mixture: Prepared and added all reagents one by one as the following order: 1 M putrescine, 1.5 M spermidine, 8.3 mM NAD, 30 mM ATP, 21.5 mM CTP, GTP and UTP, 6.8 mM CoA, 4.3 mg/mL *E. coli* tRNA, and 0.9 mg/mL folinic acid. Before adding the next reagent, ensure the last reagent was completely dissolved. The final pH should be between 7.4 and 7.6. Store at −80 °C.
- *p*PaF solution: 100 mM (Medchem Source LLP, Washington, America; Cat. no.: JA-1003).
- *E. coli* extract (Prepared in 3.1).
- The *p*PaFRS (Prepared in 3.2).
- 25 mM of DTT.
- 55 mM iodoacetamide.
- 0.1% trifluoroacetic acid in 50% acetonitrile aqueous solution.
- Sequencing grade modified trypsin (Promega, America; Cat. No. #V5117).
- Mobile phase A: 0.1% formic acid.
- Mobile phase B: 100% acetonitrile and 0.1% formic acid.

2.2. Equipment

2.2.1. Preparation of *E. coli* Extract

- 1 L flasks.
- Constant temperature shaker.
- BIOSTAT® A plus Bioreactor (Sartorius, Gottingen, Germany)
- Ultrospec 3100 pro UV/Visible spectrophotometer (Amersham, Piscataway, NJ, USA)
- Micro Refrigerated Centrifuge Model 3700 (KUBOTA, Osaka, Japan)
- Vortex-Genie 2 (Scientific Industries, Bohemia, NY, USA)
- JN-3000PLUS high press crusher (JNBIO, China)
- Spectra/Por #1 dialysis tubing, MWCO 6-8 kD (Spectrum Laboratories, Rancho Dominguez, CA, USA)
- MS-H-Pro+ magnetic stirring apparatus (DRAGONLAB, Beijing, China)

2.2.2. Preparation of *p*PaFRS

- 1 L flasks.
- Constant temperature shaker.
- JN-3000PLUS high press crusher (JNBIO)
- Ultrospec 3100 pro UV/Visible spectrophotometer (Amersham)
- Micro Refrigerated Centrifuge Model 3700 (KUBOTA)
- Vortex-Genie 2 (Scientific Industries)
- ÄKTAprime plus (GE Healthcare, Chicago, IL, USA)
- MS-H-Pro+ magnetic stirring apparatus (DRAGONLAB)
- Amicon Ultra 15 mL 10 K (Merck Millipore, Darmstadt, Germany; Cat. No. UFC901096)
- Orbital Shaker TS-2 (Kylin-Bell Lab Instruments, Haimen, China)

2.2.3. Preparation of Crude T7 RNA Polymerase

- 1 L flasks.
- Constant temperature shaker.
- Ultrospec 3100 pro UV/Visible spectrophotometer (Amersham)
- Micro Refrigerated Centrifuge Model 3700 (KUBOTA)
- Vortex-Genie 2 (Scientific Industries)
- Qsonica Q700 Ultrasonic crusher (Misonix, Farmingdale, NY, USA)
- MS-H-Pro+ magnetic stirring apparatus (DRAGONLAB)

2.2.4. Preparation of Expression Template and o-tDNA^{opt}

- Constant temperature shaker.
- Micro Refrigerated Centrifuge Model 3700 (KUBOTA)
- Vortex-Genie 2 (Scientific Industries)
- Plasmid mini kit (Omega Bio-Tek)
- C100TM Thermal Cycler (Bio-Rad)
- NanoVue Plus (GE Healthcare)

2.2.5. Synthesis and Characterization of sfGFP and sfGFP2*p*PaF

- Infinite M200 PRO Microplate reader (Tecan, Switzerland)
- Strep-Tactin affinity chromatography (IBA GmbH, Goettingen, Germany)
- Amicon Ultra 15 mL 10 K (Merck Millipore; Cat. No. UFC901096)

- Mini-PROTEAN® Tetra Cell (Bio-rad)
- SpeedVac (Thermo Fisher Scientific, Waltham, MA, USA)
- EASY-nLC 1000 system (Thermo Fisher Scientific)
- Analytical column: A home-made fused silica capillary column (75 μm ID, 150 mm length; Upchurch, Oak Harbor, WA, USA) packed with C-18 resin (300 Å, 5 μm, Varian, Lexington, MA)
- Orbitrap Fusion Tribrid mass spectrometer (Thermo Fisher Scientific, Bremen, Germany)
- Xcalibur3.0 software
- Proteome Discoverer (Version PD1.4, Thermo Fisher Scientific)

3. Procedure

3.1. Preparation of E. coli Extract (Time for completion: 2 days)

The volume of medium is one-fifth of the volume of flasks.

1. Prepared solutions used in this part.

    (1)   2 × Yeast extract-Tryptone (YT)-Phosphate (P)-Chloramphenicol (Cm) medium (10 g/L Yeast extract, 16 g/L Tryptone, 5 g/L NaCl, 40 mM K_2HPO_4, 22 mM KH_2PO_4, 34 mg/L chloramphenicol, and 1.5% agar for plate use if needed).

    (2)   S30 buffer A: 14 mM L-Glutamic acid hemimagnesium salt tetrahydrate, 60 mM Potassium L-glutamate, 50 mM Tris, pH 7.7, titrated with acetic acid. Add DTT to 2 mM just before use. Stored at 4 °C.

    (3)   S30 buffer B: 14 mM L-Glutamic acid hemimagnesium salt tetrahydrate, 60 mM Potassium L-glutamate, pH 8.2, titrated with Tris. Add DTT to 1 mM just before use. Stored at 4 °C.

2. Prepared and autoclaved 2 × YT-P medium (10 mL, 200 mL, and 4 L in the bioreactor) and S30 A buffer (500 mL) and S30 B buffer (2 L).
3. Cultivated *E. coli* Rosetta(DE3) strain on 2 × YT-P-Cm solid medium, and incubated overnight at 37 °C.
4. Selected single colony and transferred it to 10 mL liquid 2xYT-P medium in 50 mL flask with Cm and incubated overnight at 37 °C with 220 rpm.
5. Transferred 10 mL overnight culture into 200 mL fresh 2xYT-P medium in 1 L flask and continued culturing for about 2 h in the shaker.
6. When the OD600 of culture reached to 2 to 3, it was transferred into a 4-L bioreactor (Sartorius) with Cm and 400 μL antifoam. Controlled the fermentation conditions at 37 °C and 500 rpm stirring.
7. ▲ **CRITICAL STEP** When the OD600 reached to 3.5 to 4.0, harvested the cells quickly at 4 °C to obtain high activity of cell extracts.
8. ⏸ **PAUSE STEP** Washed the cell pellets with 100 mL S30 buffer A at least twice (after being washed, the pellet could be stored at 4 °C overnight or at −80 °C for a long time).
9. Re-suspended the pellets in 1 mL of S30 buffer A per gram of biomass on the ice.
10. Subjected the suspension to a high press crusher (JNBIO) twice at 15000~20000 psi.
11. Centrifuged the lysed cells at 4 °C and 13,000× *g* for at least 30 min.
12. Incubated the supernatant at 37 °C with 120 rpm for 80 min.
13. Centrifuged the extract at 4 °C and 13,000× *g* for at least 30 min.
14. Transferred the supernatant to 6–8 kDa MWCO dialysis tubing and did dialysis in 100 times volume of supernatant S30 buffer B overnight at 4 °C (or 4 h twice).
15. Re-centrifuged the extract at 4 °C with 13,000× *g* for 30 min.
16. Collected and transferred the supernatant to 1.5 mL Eppendorf tubes on ice.
17. Flash-frozen the extracts in liquid nitrogen and stored them at −80 °C.

3.2. Preparation of pPaFRS (Time for Completion: 5~6 Days from Plasmid Construction to pPaFRS Purification)

The volume of the medium is one-fifth of the volume of flasks.

1. Prepared buffers used in this part.

    (1) His-tag binding buffer: 30 mM Imidazole, 20 mM Na_3PO_4, 500 mM NaCl, titrated with phosphoric acid to pH 7.4. Stored at 4 °C.

    (2) His-tag elution buffer: 500 mM Imidazole, 20 mM Na_3PO_4, 500 mM NaCl, titrated with phosphoric acid to pH 7.4. Stored at 4 °C.

2. Amplified the pET24a vector and pPaFRS gene by Q5® High-Fidelity DNA Polymerases system.
3. Ligated the pET24a vector and pPaFRS gene with the homologous arm at 50 °C and 15 min.
4. Screened with LB-K plate (LB plate with 50 µg/mL Kanamycin) by DH5α.
5. Selected a single colony and sequenced it.
6. Transformed pET24a-6H-pPaFRS into *E. coli* BL21(DE3), spread the cell on a LB-K plate, and incubated overnight at 37 °C.
7. Picked up a single colony with a toothpick and inoculated it directly into 10 mL of LB-K medium followed by incubation for 12 h at 37 °C and 220 rpm.
8. Transferred 10 mL culture into 200 mL LB-K medium and continued culturing it overnight in a shaker.
9. Transferred the cells into 1 L fresh LB-K medium at a 5% inoculation amount.
10. When the OD600 reached 0.6–0.8, added 1 mM of Isopropyl β-D-Thiogalactoside (IPTG).
11. The cells were further cultured for 3–4 h at 37 °C. Cultivated the cells at 4 °C and 10,000 × *g*.
12. 🄿 PAUSE STEP Washed the cells with 100 mL His-tag binding buffer at least twice (after being washed, the pellets could be stored at 4 °C overnight or at −80 °C for a long time).
13. Re-suspended the pellets with suitable His-tag binding buffer, making the final optical density at a wavelength of 600 nm (OD600) between 40 and 60 on ice.
14. Subjected the suspension to high press crusher single time at 15,000~20,000 psi.
15. Centrifuged the lysate at 4 °C and 13,000 × *g* for 30 min.
16. Filtrated the lysate with 0.45 µm water filters.
17. ⚠ **CRITICAL STEP** All the lysate was loaded onto a 5 mL EzFast Ni HP column, which was connected with the ÄKTA Prime system and equilibrated with His-tag binding buffer. Then, the target pPaFRS was eluted with His-tag elution buffer and collected eluate in 1 mL fractions (all buffers should eliminate bubbles and be stored at 4 °C).
18. Placed the eluate in the 6–8 kDa MWCO dialysis tubing and dialyzed it against 50–100 volumes of sterile PBS (pH 7.4) buffer overnight.
19. Determined protein concentrations using Quick Start Bradford Protein Assay Kit. When necessary, the fractions were concentrated using Amicon Ultra centrifugal device (10 kDa).
20. Added 20% (*v*/*v*%) sucrose to fractions, and stored at −80 °C.

3.3. Preparation of Crude T7 RNA Polymerase (Time for Completion: 3 Days)

The volume of medium is one fifth of the volume of the flasks.

1. Prepared buffers used in this part (stored at 4 °C).

    (1) Lysis buffer: 50 mM NaCl, 10 mM EDTA, 10 mM K_2HPO_4, 1 mM DTT, 10 mM β-mercaptoethanol, 1 × Protease inhibitor, 5% glycerin, pH 8.0.

    (2) Dialysis buffer: 50 mM NaCl, 1 mM EDTA, 40 mM K_2HPO_4, 1 mM DTT, 20% Sucrose, pH 7.7.

2. Transformed *E. coli* BL21(DE3) with pAR1219, and incubated it on LB plate with 100 µg/mL Ampicillin (LB-A) overnight at 37 °C.
3. Selected a single colony to inoculate in 10 mL of LB-A medium and cultured cells for 12 h at 37 °C and 220 rpm.
4. Transferred 10 mL culture into 100 mL LB-A medium and continued culturing cells overnight in a shaker.
5. Transferred cells into fresh LB-A medium at a 5% inoculation amount.
6. When OD600 reached 0.6–0.8, IPTG was added to a final concentration of 0.1 mM.
7. When OD600 reached 2.0, cells were harvested for 10,000 × *g* at 4 °C.
8. ❶ **PAUSE STEP** The cells were washed with 5 mL ice-cold wash buffer per gram pellet for twice (after washing, the pellet can be stored at 4 °C overnight or at −80 °C for a long time).
9. Re-suspended in 4 mL ice-cold lysis buffer per gram cells.
10. ⚠ **CRITICAL STEP** Cells were ultrasonicated for 40 min on the ice working for two seconds and intermittent for two seconds (Kept the sample on the ice to maintain the activity).
11. Cells were centrifuged at 13,000 × *g* and 4 °C for 20 min and discarded cell pellet.
12. Dialyzed (6–8 kDa) twice in the dialysis buffer (100 times volume of samples) at 4 °C overnight. The suspension was centrifuged at 10,000× *g* for 30 min at 4 °C and the pellet was discarded.
13. The crude T7 RNA polymerase was flash-frozen in liquid nitrogen, and stored at −80 °C until use.

3.4. Preparation of Expression Template and o-tDNA^{opt} (Time for Completion: 2 Days)

1. Made pET23a-sfGFP-StrepII and pET23a-sfGFP(2TAG)-StrepII plasmids self-ligated by homology arms at 50 °C at least 15 min.
2. Added the ligation system into 100 µL DH5α competent cells and put them on ice for 30 min.
3. Added LB medium about 400 µL and cultured them at 37 °C and 220 rpm about 30 min.
4. Coated about 100 µL suspension on the LB-A plate.
5. Picked up two single colonies to sequence.
6. Selected the correct strain and extracted the plasmid with QIAGEN Plasmid Maxi Kit.
7. Ligated the o-tDNA^{opt} and the pET23a vector genes at 50 °C at least 15 min.
8. Screening method was the same as 3.4 2–5.
9. Extracted the pET23a o-tDNA^{opt} with Plasmid Mini Kit as an o-tDNA^{opt} amplification template.
10. Amplified o-tDNA gene with Pfu polymerase.
11. Purified o-tDNA gene by ethanol precipitation.
12. The o-tDNA^{opt} was diluted with MiliQ water to 2 mg/mL and stored at −20 °C.

3.5. Synthesis and Characterization of sfGFP and sfGFP2pPaF (Time for Completion: 3~4 Days)

1. Thawed the CFPS, OTS components, and expression templates on the ice.
2. The standard 20 µL cell-free reaction mixture consisted of the followings in Table 1.
3. Mixed the mixture and reacted at 30 °C for 16 h.
4. Five-microliter samples from each reaction mixture were diluted with 195 µL ddH$_2$O, and the fluorescence intensity of these diluted samples was measured with F485 excitation and F535 emission filters using Microplate reader.
5. Preparation of samples for mass spectrometry:

   (1) Purified sfGFP and sfGFP2pPaF produced in CFPS with a C-terminal Strep-tag via Strep-Tactin affinity chromatography according to the manufacturer instructions.
   (2) Dialyzed and concentrated the fractions obtained from Strep-Tactin affinity chromatography with PBS (pH 7.4) buffer at 4 °C.

(3)   Determined protein concentrations using Quick Start Bradford Protein Assay Kit.

(4)   Analyzed the samples by sodium dodecyl sulfate polyacrylamide gel electrophoresis (SDS-PAGE).

(5)   Extracted interested bands from the gel.

(6)   Reduced with 25 mM of DTT and alkylated with 55 mM iodoacetamide and digested in gel with sequencing grade modified trypsin at 37 °C overnight.

(7)   Extracted peptides twice with 0.1% trifluoroacetic acid in 50% acetonitrile aqueous solution for 30 min and then dried in a SpeedVac.

(8)   Redissolved peptides in 25 µL 0.1% trifluoroacetic acid.

(9)   Analyzed 6 µL of extracted peptides by Thermo orbitrap fusion.

(10)   LC-MS/MS analysis of samples was performed at the Center of Biomedical Analysis of Tsinghua University.

Table 1. Components of standard cell-free protein synthesis (CFPS) system.

Components	Volume (µL)
10 × salt	2
PEP	1.6
Mg^{2+}	0.4
Extract	5
T7 polymerase	0.2
NTPs mixture	0.8
Amino acids mixture	0.8
Plasmid template	1 (300 ng)
o-tDNA	1 (100 ng/µL)
pPaFRS	1 (5 mM)
ddH$_2$O	6.12
Total	20

PEP: Phosphoenolpyruvic acid; pPaFRS: p-propargyloxyphenylalanine.

4. Expected Results

4.1. Preparation of pPaFRS and o-tDNA^{opt}

Good preparation of pPaFRS and o-tDNA^{opt} is the first step to successful incorporation of UNAAs. pPaFRS proteins were purified by His-tag affinity chromatography, concentrated to 10 mg/mL by ultrafiltration tube, and stored with 20% sucrose. From the SDS-PAGE result (Figure 2a), the only protein band indicated successful purification. The o-tDNA genes were amplified by PCR and purified by ethanol precipitation. Analyzed by 1% agarose electrophoresis (Figure 2b), the o-tDNA gene was the only product. Thus, ethanol precipitation used in this study was suitable for o-tDNA purification, and this product was favorable for subsequent CFPS reaction.

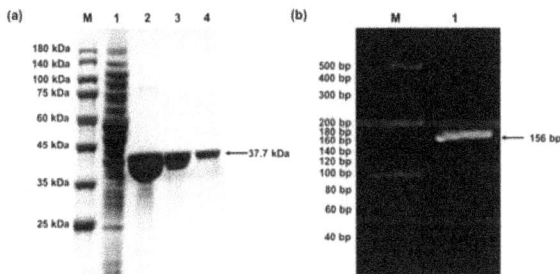

Figure 2. (a) SDS-PAGE result of pPaFRS. M: protein marker (PM2510, Transgen, China); 1: Cell lysate of induced BL21(DE3); 2, 3, and 4: different loading amounts of pPaFRS. (b) The o-tDNA gene by Polymerase Chain Reaction (PCR) amplification. M: 20 bp DNA Ladder (TaKaRa, Japan; Cat. No. #3420A). 1: The product of the o-tDNA gene by PCR.

4.2. Synthesis and Characterization of sfGFP and sfGFP2pPaF

4.2.1. Synthesis and Characterization of sfGFP and sfGFP2pPaF

To screen the optimum conditions, reactions were conducted at 25 °C, 30 °C, and 37 °C, and then, the fluorescence intensity was measured at different time points (Figure 3a). According to the results, the reaction proceeded fastest at 37 °C; however, the fluorescence intensity was the weakest among the three different temperatures. At 25 °C, the reaction kept steady after 22 h, and the fluorescence intensity was weaker than at 30 °C. The fluorescence intensity was the strongest at 30 °C and kept steady after 13 h. Therefore, the optimum reaction temperature was 30 °C.

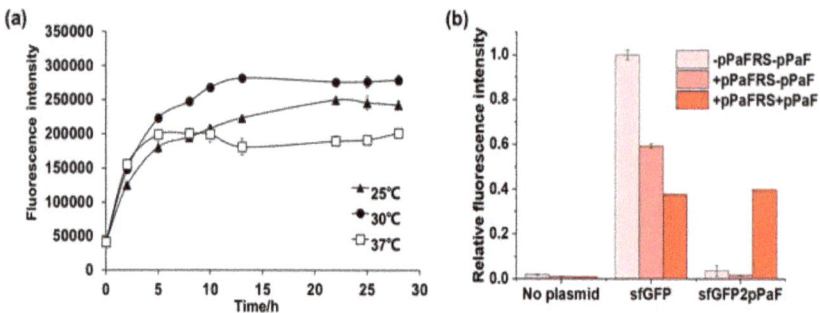

Figure 3. (a) Time course of sfGFP2pPaF synthesis catalyzed by purified OTS. Two independent batch CFPS reactions (n = 2) were performed at 25 °C, 30 °C and 37 °C for each point over 24 h. (b). Synthesis of sfGFP2pPaF at 30 °C for 16 h.

To further investigate the incorporation efficiency, reactions were conducted at 30 °C in 16 h. The experiment was designed into three groups. The first group was without pPaFRS and pPaF. The second was only pPaFRS without pPaF. The last group was added with both pPaFRS and pPaF. When the pET23a-sfGFP(2TAG)-StrepII plasmid was added into cell-free reactions, the fluorescence of the last group with both pPaFRS and pPaF should increase obviously; however, the other two groups should present weaker fluorescence. As shown in Figure 3b, this indicated that the addition of OTS components could reduce the expression of sfGFP by 40% to 60%, which suggested that the unnatural OTS components indeed were harmful to the natural system. It could be observed that more rigorous incorporation has been achieved because the fluorescence intensity of sfGFP2pPaF experimental group was far stronger than its control group. In brief, pPaF could be efficiently incorporated by this protocol.

4.2.2. Characterization of sfGFP and sfGFP2pPaF

The unnatural proteins produced in CFPS system need to be further characterized. Wild-type sfGFP and unnatural sfGFP2pPaF proteins were first purified by Strep-Tactin affinity chromatography. Fractions obtained from Strep-Tactin affinity chromatography were analyzed by SDS-PAGE and LC-MS/MS. The SDS-PAGE results (Figure 4a) revealed that the target protein has only one band, which meant successful protein purification. As shown in Figure 4b, mass spectrometry detected modification of an alkynyl group at the original second position and a molecular weight increase of 54.01063. The molecular weight of pPaF is 54.01063 higher than that of Phe. The LC-MS/MS analysis confirmed that pPaF had been successfully incorporated into the sfGFP protein.

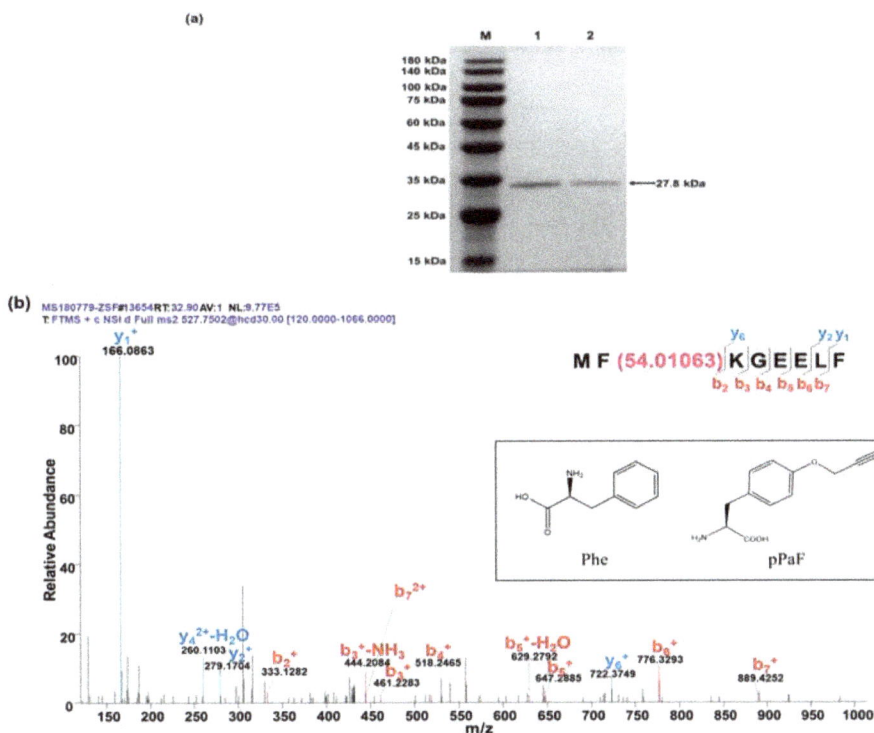

Figure 4. (a) Purified protein with C-terminal Strep-tag. 1: wild-type sfGFP; 2: sfGFP2*p*PaF. (b) LC-MS/MS analysis of sfGFP2*p*PaF.

5. Conclusions

This article described a method of synthesizing unnatural proteins that contained UNAAs at a specific single position. Compared with previous methods, this method used OTS components separately and achieved more controllable and efficient incorporation of UNAAs. Therefore, the whole experimental cycle could be shortened and need not prepare cell extracts expressing different OTSs. In this strategy, o-tRNA was added indirectly by o-tDNA transcription promoted by T7 promoter, which avoided redundant transcription in vitro. Consequently, this method is quick, convenient, and highly efficient, and it could be developed as the standard protocol for unnatural protein synthesis.

Supplementary Materials: The following are available online at http://www.mdpi.com/2409-9279/2/1/16/s1, Figure S1: Nucleotide and amino acid sequences of *p*PaFRS, Figure S2: Nucleotide and amino acid sequences of original sfGFP, Figure S3: Nucleotide and amino acid sequences of mutational sfGFP2*p*PaF, Figure S4: Nucleotide sequence of o-tDNA, Table S1: Primers used in this article.

Author Contributions: Conceptualization and methodology, all authors; validation and formal analysis, W.G.; writing—original draft preparation, W.G.; writing—review and editing, N.B. and Y.L.; supervision, Y.L.; funding acquisition, Y.L.

Funding: This work was supported by the National Natural Science Foundation of China, grant number 21706144 and 21878173.

Acknowledgments: Many thanks to Chongchong Zhao (the Center of Biomedical Analysis, Tsinghua University) for the assistance with LC-MS/MS analysis.

Conflicts of Interest: The authors declare no conflicts of interest.

References

1. Soye, B.J.D.; Patel, J.R.; Isaacs, F.J.; Jewett, M.C. Repurposing the translation apparatus for synthetic biology. *Curr. Opin. Chem. Biol.* **2015**, *28*, 83–90. [CrossRef] [PubMed]
2. Johnson, J.A.; Lu, Y.Y.; Van Deventer, J.A.; Tirrell, D.A. Residue-specific incorporation of non-canonical amino acids into proteins: Recent developments and applications. *Curr. Opin. Chem. Biol.* **2010**, *14*, 774–780. [CrossRef] [PubMed]
3. Wang, L. Genetically encoding new bioreactivity. *New Biotechnol.* **2017**, *38*, 16–25. [CrossRef] [PubMed]
4. Leisle, L.; Valiyaveetil, F.; Mehl, R.A.; Ahern, C.A. Incorporation of non-canonical amino acids. *Adv. Exp. Med. Biol.* **2015**, *869*, 119–151. [PubMed]
5. Quast, R.B.; Mrusek, D.; Hoffmeister, C.; Sonnabend, A.; Kubick, S. Cotranslational incorporation of non-standard amino acids using cell-free protein synthesis. *FEBS Lett.* **2015**, *589*, 1703–1712. [CrossRef] [PubMed]
6. Gan, R.; Perez, J.G.; Carlson, E.D.; Ntai, I.; Isaacs, F.J.; Kelleher, N.L.; Jewett, M.C. Translation system engineering in *Escherichia coli* enhances non-canonical amino acid incorporation into proteins. *Biotechnol. Bioeng.* **2017**, *114*, 1074–1086. [CrossRef] [PubMed]
7. Hong, S.H.; Kwon, Y.-C.; Jewett, M.C. Non-standard amino acid incorporation into proteins using *Escherichia coli* cell-free protein synthesis. *Front. Chem.* **2014**, *2*, 34. [CrossRef] [PubMed]
8. O'Donoghue, P.; Ling, J.; Wang, Y.-S.; Söll, D. Upgrading protein synthesis for synthetic biology. *Nat. Chem. Biol.* **2013**, *9*, 594. [CrossRef] [PubMed]
9. Bröcker, M.J.; Ho, J.M.L.; Church, G.M.; Söll, D.; O'Donoghue, P. Recoding the genetic code with selenocysteine. *Angew. Chem. Int. Ed.* **2014**, *53*, 319–323. [CrossRef] [PubMed]
10. Summerer, D.; Chen, S.; Wu, N.; Deiters, A.; Chin, J.W.; Schultz, P.G. A genetically encoded fluorescent amino acid. *Proc. Natl. Acad. Sci. USA* **2006**, *103*, 9785–9789. [CrossRef] [PubMed]
11. Oza, J.P.; Aerni, H.R.; Pirman, N.L.; Barber, K.W.; ter Haar, C.M.; Rogulina, S.; Amrofell, M.B.; Isaacs, F.J.; Rinehart, J.; Jewett, M.C. Robust production of recombinant phosphoproteins using cell-free protein synthesis. *Nat. Commun.* **2015**, *6*, 8168. [CrossRef] [PubMed]
12. Albayrak, C.; Swartz, J.R. Direct polymerization of proteins. *ACS Synth. Biol.* **2014**, *3*, 353–362. [CrossRef] [PubMed]
13. Cho, H.; Daniel, T.; Buechler, Y.J.; Litzinger, D.C.; Maio, Z.; Putnam, A.-M.H.; Kraynov, V.S.; Sim, B.-C.; Bussell, S.; Javahishvili, T.; et al. Optimized clinical performance of growth hormone with an expanded genetic code. *Proc. Natl. Acad. Sci. USA* **2011**, *108*, 9060–9065. [CrossRef] [PubMed]
14. Bundy, B.C.; Swartz, J.R. Site-specific incorporation of p-propargyloxyphenylalanine in a cell-free environment for direct protein-protein click conjugation. *Bioconj. Chem.* **2010**, *21*, 255–263. [CrossRef] [PubMed]
15. Lu, Y. Cell-free synthetic biology: Engineering in an open world. *Synth. Syst. Biotechnol.* **2017**, *2*, 23–27. [CrossRef] [PubMed]
16. Carlson, E.D.; Gan, R.; Hodgman, C.E.; Jewett, M.C. Cell-free protein synthesis: Applications come of age. *Biotechnol. Adv.* **2012**, *30*, 1185–1194. [CrossRef] [PubMed]
17. Caschera, F.; Noireaux, V. Synthesis of 2.3 mg/mL of protein with an all *Escherichia coli* cell-free transcription-translation system. *Biochimie* **2014**, *99*, 162–168. [CrossRef] [PubMed]
18. Deiters, A.; Schultz, P.G. In vivo incorporation of an alkyne into proteins in *Escherichia coli*. *Bioorg. Med. Chem. Lett.* **2005**, *15*, 1521–1524. [CrossRef] [PubMed]
19. Chin, J.W.; Santoro, S.W.; Martin, A.B.; King, D.S.; Wang, L.; Schultz, P.G. Addition of p-azido-l-phenylalanine to the genetic code of *Escherichia coli*. *J. Am. Chem. Soc.* **2002**, *124*, 9026–9027. [CrossRef] [PubMed]
20. Albayrak, C.; Swartz, J.R. Cell-free co-production of an orthogonal transfer RNA activates efficient site-specific non-natural amino acid incorporation. *Nucleic Acids Res.* **2013**, *41*, 5949–5963. [CrossRef] [PubMed]

methods and protocols

MDPI

Protocol

A Crude Extract Preparation and Optimization from a Genomically Engineered *Escherichia coli* for the Cell-Free Protein Synthesis System: Practical Laboratory Guideline

Jeehye Kim [1,†], Caroline E. Copeland [1,†], Sahana R. Padumane [1] and Yong-Chan Kwon [1,2,*]

[1] Department of Biological and Agricultural Engineering, Louisiana State University, Baton Rouge, LA 70803, USA
[2] Louisiana State University Agricultural Center, Baton Rouge, LA 70803, USA
* Correspondence: yckwon@lsu.edu; Tel: +1-225-578-4325
† These authors contributed equally to this work.

Received: 2 July 2019; Accepted: 7 August 2019; Published: 9 August 2019

Abstract: With the advancement of synthetic biology, the cell-free protein synthesis (CFPS) system has been receiving the spotlight as a versatile toolkit for engineering natural and unnatural biological systems. The CFPS system reassembles the materials necessary for transcription and translation and recreates the in vitro protein synthesis environment by escaping a physical living boundary. The cell extract plays an essential role in this in vitro format. Here, we propose a practical protocol and method for *Escherichia coli*-derived cell extract preparation and optimization, which can be easily applied to both commercially available and genomically engineered *E. coli* strains. The protocol includes: (1) The preparation step for cell growth and harvest, (2) the thorough step-by-step procedures for *E. coli* cell extract preparation including the cell wash and lysis, centrifugation, runoff reaction, and dialysis, (3) the preparation for the CFPS reaction components and, (4) the quantification of cell extract and cell-free synthesized protein. We anticipate that the protocol in this research will provide a simple preparation and optimization procedure of a highly active *E. coli* cell extract.

Keywords: cell-free protein synthesis; *E. coli* crude extract preparation; genomically engineered *E. coli*; sonication

1. Introduction

The technology involving the disruption of bacterial cells and collection of ribosomes for synthesizing proteins was first introduced when the fraction of ribosomes was identified as the core of the protein synthesis machinery of the cells [1]. The cell-free protein synthesis (CFPS) system has been developed for exclusive protein synthesis utilizing active ribosomes and other cellular machinery outside of the living cell [2,3]. Recent progress of synthetic biology highlights this versatile system as an essential toolkit for exploring and maneuvering complex cellular processes to accelerate technology advances [4,5]. The CFPS system offers many advantages over a cell-based system, such as ease of manipulating biochemical pathways [6,7], higher tolerance on chemicals and toxic compounds [8,9], utilization of PCR amplified linear template allowing for high-throughput preparation and protein synthesis of the gene of interest and breadboarding synthetic biological circuit [10–13], and the capability of highly efficient non-standard amino acid incorporation [14,15]. In addition, the CFPS system allows the benefits of unprecedented logistics along with freeze-drying paper-based format [16].

Since the cell extract carries most of the cellular machinery, its preparation is considered as the first important step for building a highly productive CFPS system. Many studies streamlined the overall procedure for cell extract preparation to improve overall extract performance in CFPS system [17–21].

Although recent progress in the preparation of *Escherichia coli* cell extract has resulted in an increase in protein productivity up to 1–1.5 mg/mL in a single batch cell-free system [22], the total protein yields are varied from strain to strain due to the necessary variations, dictated by the strain, of the three major cell extract preparation stages: pre-lysis, lysis, and post-lysis. The preparation step of the cell extract is crucial, as it contains key components for synthesizing proteins, so it is important to practice the optimized cell-free extract method for each *E. coli* strain for high protein yield. For example, Kim et al. [19] demonstrated that *E. coli* strain BL21(DE3)-derived cell extract prepared in a simpler procedure were more effective than the cell extracts prepared by the conventional method described by Pratt et al. [23]. However, the modified procedure was not as productive for the traditional host organism, *E. coli* A19 derived extract [19]. This evidence indicated that different preparation conditions are required depending on the *E. coli* strain of choice to maximize cell extract performance. However, the recent study from Kwon and Jewett was the first to introduce an optimized systemic cell extract preparation process for the non-commercial engineered *E. coli* strain K12 MG1655 (C495) which greatly advanced the potential for use in future biomedical/industrial applications [22]. In addition, the systemically optimized CFPS is inspiring novel ways to utilize the cell extracts from engineered *E. coli* strains for applications involving unnatural amino acid incorporation [15], patient-specific therapeutic vaccines [24], anticancer protein production [25], and more.

Here we describe a procedure for cell extract preparation step-by-step for the genomically engineered *E. coli* strain Δ*prfA* Δ*endA* Δ*rne* [15] to generate the optimal cell extract with the maximum protein productivity. This protocol also can be applied to other strains with slight modification. The protocol aims to clarify which processing variables are the most critical for the cell extract performance during CFPS and how the processing condition can be optimized for different *E. coli* strains.

2. Experimental Design

The protocol in this study was designed to obtain the optimal cell extract that can maximize protein production yield during the cell-free protein synthesis. As illustrated in Figure 1, the streamlined processing steps were considered to be the most important parameters that influence the overall activity of cell extract: Pre-lysis (culture and harvest), lysis, and post-lysis (run-off reaction and dialysis) steps. Briefly, the procedures consist of three parts: (1) Tuning the cell culture and harvest time to obtain the most actively growing cells, (2) characterizing cell lysis condition by evaluating the relative protein production yield of each cell extract across sonication energy input and processing volume, and (3) an optional step of conducting a run-off reaction and dialysis for clearing out the cell extract [22,26]. This protocol will provide a practical guideline to produce a highly active cell extract by describing the overall extract preparation processes in detail for a better understanding of the CFPS system.

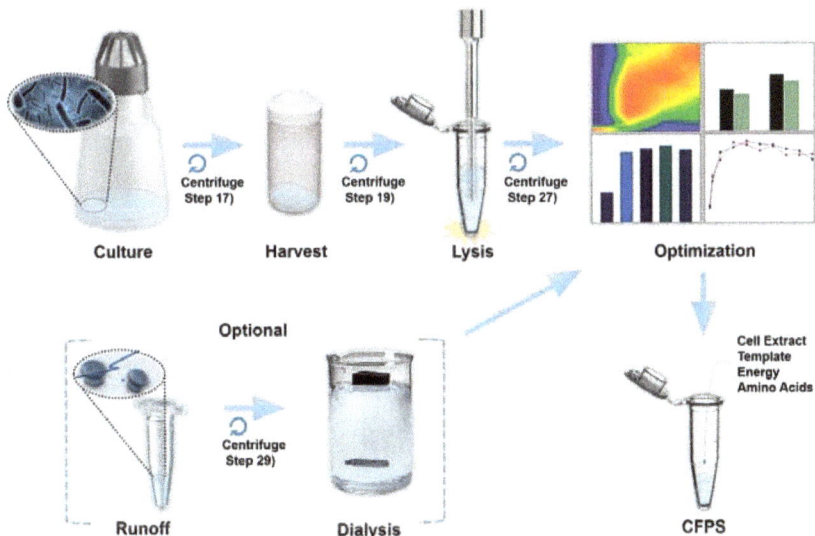

Figure 1. A workflow of the cell extract preparation and optimization.

2.1. Materials

2.1.1. E. coli Cell Preparation: Culture Media (2xYTPG and LB) and LB Agar Plate

- Genomically engineered *E. coli* strain K12 MG1655 ΔprfA ΔendA Δrne [15]
- Ampicillin sodium salt (Fisher Bioreagents, Fair Lawn, NJ, USA)
- Bacto™ Tryptone (BD Biosciences, San Jose, CA, USA)
- Bacto™ yeast extract (BD Biosciences)
- Bacto™ agar (BD Biosciences)
- Glucose (Fisher Bioreagents)
- Sodium chloride (Fisher Bioreagents)
- Potassium phosphate dibasic (K_2HPO_4) (Fisher Bioreagents)
- Potassium phosphate monobasic (KH_2PO_4) (Fisher Bioreagents)
- Potassium hydroxide (Fisher Bioreagents)

2.1.2. Cell Disruption and Extract Preparation: Buffer A and Dialysis

- 1,4-Dithio-DL-threitol (DTT) (Sigma-Aldrich, St. Louis, MO, USA)
- Potassium acetate (Fisher Bioreagents)
- Magnesium acetate tetrahydrate (Fisher Bioreagents)
- Tris base (Fisher Bioreagents)
- Potassium hydroxide (Fisher Bioreagents)
- Slide-A-Lyzer™ G2 dialysis cassette, 10K MWCO, 3 mL (Thermo Fisher Scientific, Waltham, MA, USA) for dialysis (if necessary)

2.1.3. Cell-Free Protein Synthesis

- Magnesium glutamate (Sigma-Aldrich)
- Ammonium glutamate (MP Biomedicals, Santa Ana, CA, USA)
- Potassium glutamate (Sigma-Aldrich)
- Adenosine triphosphate (ATP) (Alfa Aesar, Haverhill, MA, USA)

- Cytidine triphosphate (CTP) (Alfa Aesar)
- Uridine triphosphate (UTP) (Alfa Aesar)
- Guanosine triphosphate (GTP) (Sigma-Aldrich)
- L-5-formyl-5, 6, 7, 8-tetrahydrofolic acid (Folinic acid) (Alfa Aesar)
- *E. coli* total tRNA mixture (from strain MRE600) (Roche Applied Science, Indianapolis, IN, USA)
- Phosphoenolpyruvate (PEP) (Alfa Aesar)
- 20 amino acids (Alfa Aesar)
- Nicotinamide adenine dinucleotide (NAD) (Thermo Fisher Scientific)
- Coenzyme-A (CoA) (Sigma-Aldrich)
- Potassium oxalate (Oxalic acid) (Alfa Aesar)
- Putrescine (Thermo Fisher Scientific)
- Spermidine (Thermo Fisher Scientific)
- HEPES (Thermo Fisher Scientific)
- Phusion® high-fidelity DNA polymerase (New England Biolabs, Ipswich, MA, USA) for the preparation of PCR amplified linear template

2.1.4. Protein Quantification

- NuPAGE® 4–12% bis-tris gel (Thermo Fisher Scientific)
- NuPAGE® LDS sample buffer (Thermo Fisher Scientific)
- NuPAGE® MES SDS running buffer (20×) (Thermo Fisher Scientific)
- SimplyBlue™ SafeStain (Thermo Fisher Scientific)
- Coomassie Blue assay reagent (Thermo Fisher Scientific)
- 1,4-Dithio-DL-threitol (DTT) (Sigma-Aldrich)

2.2. Equipment

- 2.5 L baffled Tunair shake flasks (IBI Scientific, Peosta, IA, USA)
- 300 mL baffled Tunair shake flasks (IBI Scientific)
- New Brunswick™ Innova® 42 incubator shaker (Eppendorf, Hamburg, Germany)
- Genesys™ 6 UV-Vis spectrophotometer (Thermo Fisher Scientific)
- Sorvall Legend X1 Sorvall Legend X1R centrifuge (Thermo Fisher Scientific, USA)
- TX-400 swinging bucket rotor (Thermo Fisher Scientific)
- Round Buckets for TX-400 rotor (Thermo Fisher Scientific)
- Fiberlite™ F15-8 × 50cy fixed angle rotor (Thermo Fisher Scientific)
- Thermo Scientific set of four adapters for 15 mL Conical Tube (Thermo Fisher Scientific)
- MicroClick 30 × 2 fixed angle microtube rotor (Thermo Fisher Scientific)
- Q125 Sonicator with 1/8" (3 mm) diameter probe (Qsonica, Newtown, CT, USA)
- Synergy™ HTX multi-mode microplate reader (BioTek, Winooski, VT, USA)
- Genesys™ 6 UV-Vis spectrophotometer (Thermo Fisher Scientific)
- Fisherbrand™ accuSpin™ micro 17 microcentrifuges (Thermo Fisher Scientific)
- PowerPac™ basic power supply (Bio-Rad, Hercules, CA, USA)
- Mini gel tank (Invitrogen, Carlsbad, CA, USA)
- Direct-Q3® UV water purification system (Millipore, Burlington, MA, USA)
- Scotsman flake ice maker (CurranTaylor, Canonsburg, PA, USA)

3. Procedure

3.1. Cell Extract Preparation. Time for Completion: Four Days

3.1.1. Day 1

(1)   LB media: Dissolve 10 g/L of tryptone, 5 g/L of yeast extract, 10 g/L of sodium chloride in Milli-Q water and sterilize at 121 °C for 30 min.

(2)   LB-agar plate: Dissolve 10 g/L of tryptone, 5 g/L of yeast extract, 10 g/L of sodium chloride, and 15 g/L of bacto-agar in Milli-Q water and sterilize 121 °C for 30 min. Place the container in 55–65 °C water bath to cool the solution. Add appropriate antibiotics (if necessary), mix well, and solidify in Petri dishes (20 mL each).

(3)   Streak LB-agar plate with *E. coli* K12 MG1655 Δ*prfA* Δ*endA* Δ*rne* and incubate overnight (16–20 h) at 34 °C.

(4)   Put 1 L of Milli-Q water and centrifuge rotors in 4 °C.

(5)   Prepare three buffer A stock solutions (100x) in separate bottles (500 mL each), (a) 1 M Tris-acetate (pH 8.2): Dissolve 60.57 g of Tris base in 500 mL of Milli-Q water and adjust pH to 8.2 with 5 N potassium hydroxide, (b) 1.4 M Magnesium acetate: dissolve 107.23 g of magnesium acetate tetrahydrate in 500 mL of Milli-Q water, (c) 6 M potassium acetate: Dissolve 294.42 g of potassium acetate in 500 mL of Milli-Q water. All solutions are filtrated by passing them through a 0.2 µm filter unit.

3.1.2. Day 2

(1)   Pick a single bacterial colony (2–3 mm in diameter) from the plate and transfer the colony into 30 mL of LB media in a 125 mL baffled flask with appropriate antibiotics.

(2)   Incubate the culture for overnight (8–12 h) at 34 °C with vigorous shaking at 250 rpm.

(3)   Glucose solution: 18 g/L of glucose (0.4 M) in Milli-Q water.

(4)   2xYTP media: dissolve 16 g of tryptone, 10 g of yeast extract, 5 g of sodium chloride, 7 g of potassium phosphate dibasic, and 3 g of potassium phosphate monobasic in 500 mL of Milli-Q water in 1 L beaker.

(5)   Adjust pH with 5 N potassium hydroxide until it reaches 7.2 and then add Milli-Q water until it reaches 750 mL.

(6)   Transfer 750 mL 2xYTP (pH 7.2) to 2.5 L baffled Tunair shake flask, close the cap, and warp with aluminum foil.

(7)   ⚠ **CRITICAL STEP** 2xYTP media and glucose solution are prepared in separate containers and sterilize at 121 °C for 30 min separately.

3.1.3. Day 3

(8)   Add 250 mL of sterilized 0.4 M glucose solution into 750 mL of sterilized 2xYTP media near the flame or in the biosafety cabinet right before inoculation. Shake well. The total 2xYTPG media volume is 1 L. Antibiotics are not included.

(9)   Transfer 10 mL of overnight culture (1:100 ratio) to 1 L main culture media (step (11)) near the flame or in the biosafety cabinet.

(10)   ⚠ **CRITICAL STEP** Monitor the cell growth rate (OD_{600}) including initial inoculum by spectrophotometer until OD_{600} reaches to the mid-exponential growth phase. Calculate cell doubling time (T_d). Optimal doubling time is 35.6 ± 0.3 [15]

(11)   ⚠ **CRITICAL STEP** From here, the cells need to handle at 4 °C or below. Put the centrifuge bottles in ice before harvest. Weigh empty 50 mL conical tubes and mark weight (g) on the tube and put in ice. Turn on the centrifuge and set the temperature at 4 °C.

(12) Harvest the cell when the growth curve reaches at a mid-exponential phase by centrifugation (TX-400 swinging bucket rotor) at 5000 RCF at 4 °C for 15 min and then discard the supernatants.

(13) Prepare buffer A solution: Combine 10 mL of each stock solutions with 970 mL of 4 °C chilled Milli-Q water (from Day 1) and add 1 mL of 1 M DTT (Table 1). Mix well and put in ice.

Table 1. The components for 1 L of buffer A.

Reagents	Mixing Volume (mL)
1 M Tris-OAc, pH 8.2	10
1.4 M Mg(OAc)$_2$	10
6 M KOAc	10
1 M DTT	1
Chilled Milli-Q water	970

(14) Cell washing: transfer the harvested cells to 50 mL conical tube and add chilled buffer A solution up to 40 mL. Resuspend the cells by shaking or vortexing. ▲ **CRITICAL STEP** Maintain low temperature during washes. Centrifugation (Fiberlite™ F15-8 × 50cy fixed angle rotor) at 5000 RCF at 4 °C for 10 min and then discard the supernatants. Repeat three times.

(15) ▲ **CRITICAL STEP** Weigh wet cell weight (g) and marked on the conical tube.

(16) ⓪ **PAUSE STEP** Flash freeze the pelleted cells in liquid nitrogen and stored at −80 °C until the next step.

3.1.4. Day 4 (or Continue Day 3)

(17) Take out frozen cells from −80 °C and thaw in ice.

(18) Resuspend the cells in 1 mL of buffer A per 1 g of wet cell mass.

(19) ▲ **CRITICAL STEP** Transfer the cell suspension into 1.5 mL microtube in an ice-water container to minimize heat damage during sonication.

(20) The cell suspension is lysed using sonicator at a frequency of 20 kHz and 50% amplitude. Different levels of energy input (Joules) and sample volumes are applied during sonication to obtain the highest protein production yield for CFPS reaction.

(21) ▲ **CRITICAL STEP** Add DTT into cell lysate (3 µL per 1 mL lysate), invert several times quickly, and place the microtube in the ice-water bucket until the next step.

(22) The lysate is centrifuged once at 12,000 RCF at 4 °C for 10 min.

(23) ⓪ **PAUSE STEP** Transfer the supernatant (crude cell extract) to a fresh microtube, flash freeze in liquid nitrogen, and store at −80 °C until use.

(24) **OPTIONAL STEP** The runoff reaction (pre-incubation at 37 °C at 250 rpm) is performed in different reaction time (0, 20, 40, 60, and 80 min). After runoff reaction, clear the cell extract by centrifugation at 10,000 RCF at 4 °C for 10 min.

(25) **OPTIONAL STEP** The cell extract can be dialyzed if necessary. Cell extract transfer to dialysis cassette (Slide-A-Lyzer™ G2 dialysis cassette, 10K MWCO) and dialyze in chilled buffer A with stirring at 4 °C. Buffer A is exchanged every 45 min four times.

3.2. Cell-Free Protein Synthesis. Time for Completion: One to Two Days

(1) The standard CFPS reaction mixture (15 µL) consists of following components: 12 mM magnesium glutamate, 10 mM ammonium glutamate, 130 mM potassium glutamate, 1.2 mM ATP, 0.85 mM each of GTP, UTP, and CTP, 34.0 µg/mL folinic acid, 171 µg/mL *E. coli* total tRNA (strain MRE600), 2 mM each of 20 amino acids, 33 mM phosphoenolpyruvate (PEP), 0.33 mM nicotinamide adenine dinucleotide (NAD), 0.27 mM coenzyme-A (CoA), 1.5 mM spermidine, 1 mM putrescine, 4 mM sodium oxalate, 57 mm HEPES-KOH (pH 7.5), 100 µg/mL T7 RNA polymerase, 13.3 ug/mL pJL1-sfGFP plasmid and 27% v/v of cell extract. All components store at −80 °C until use.

(2) Place dry heat block (filled with water) in the incubator. Set temperature at 30 °C or 37 °C.

(3) ⚠ **CRITICAL STEP** Take out all CFPS components from the freezer and thaw in ice. Prepare fresh microtubes for the CFPS reaction in ice.

(4) Vortex all CFPS reaction components well except for T7 RNA polymerase, cell extract, and plasmid. To mix T7 RNA polymerase, cell extract, and plasmid gently flick the tube.

(5) ⚠ **CRITICAL STEP** Mix all reaction components in a fresh microtube and pipetting up and down to mix homogeneously and minimize bubbles.

(6) Briefly spin down the microtubes.

(7) Put the microtubes in the dry heat block and incubate for the desired time period (20 h for plasmid, 4 h for PCR amplified linear template).

(8) Put all CFPS reaction components back in the freezer for next use.

3.3. Protein Quantification

3.3.1. Superfolder Green Fluorescent Protein (sfGFP)—Fluorescence. Time for Completion: 30 min

(1) After the CFPS reaction, briefly vortex microtube.

(2) Aliquot 48 µL of Milli-Q water in 96-well flat bottom black half-well microplate and add 2 µL of cell-free synthesized protein into each well to bring the total volume to 50 µL.

(3) Place the microplate in the plate reader and set it to 15 s orbital shaking (fast) and measure the fluorescence. Excitation and emission wavelengths are 485 and 528 nm, respectively.

3.3.2. SDS-PAGE. Time for Completion: 3 h

(1) Set the dry heat block at 80 °C.

(2) Mix 5 µL of the cell-free synthesized sample, 5 µL of 4x sample loading dye, and 10 µL of 200 mM DTT solution to bring the total volume to 20 µL.

(3) Vortex well and incubate 5 min in 80 °C dry heat block.

(4) Prepare 4–12% SDS-PAGE, SDS-PAGE loading buffer, and gel tank. Clear out the gel wells by injection of the loading buffer.

(5) After 5 min of protein denaturation, spin down and vortex the samples.

(6) Load 10 µL of samples mixture to each well. For the size determination, 5 µL of pre-stained protein ladder is loaded to the first well.

(7) Run the gel electrophoresis with the voltage 150 V for one hour.

(8) Cell-free synthesized proteins are visualized by Coomassie blue staining.

3.3.3. Total Protein—Bradford Assay. Time for Completion: 30 min

(1) Prepare the BSA dilutes (2 mg/mL to 25 µg/mL) for standards and use ultra-pure water as a blank (0 µg/mL).

(2) The original cell extracts are pre-diluted 40-fold in ultra-pure water (i.e., 5 µL of cell extract plus 195 µL of water) and then use as test samples for the protein assay.

(3) Mix the protein-dye solution with standards or samples, incubate for 10 min at room temperature, and measure the absorbance at 595 nm.

(4) The original (undiluted) concentration is determined by multiplying 40 to the concentration of the diluted test sample.

(5) The amount of total protein in the cell extracts are ranged from 15.4 to 54.8 mg/mL depends on the sonication energy input and volume.

4. Expected Results

For many years, the procedure described by Pratt [23] has been considered as the standard method for preparation of *E. coli* cell extract. However, this standard process requires a lengthy preparation time in combination with multiple steps, which can cause complications when scaling up the cell-extract resulting in the inconsistent protein productivity [27]. Over the decades, studies on cell extract preparation have progressed and accomplished to show strong evidence that the systemic optimization of key parameters can significantly increase the protein production yields through the cell-free protein synthesis reaction.

4.1. Pre-Lysis Procedure

Manipulating bacterial cell culture condition influences the cell's physiological state and cellular composition over the course of its growth. Several chemical and physical parameters are considered as essential factors which can enhance metabolic properties during the culture, including culture media, temperature, pH, and oxygen. It is important to pay attention to the method that the cells are grown and harvested because these conditions directly impact the protein synthesis performance of the resulting cell extract during the CFPS reaction. For example, using the cell extract that was cultured in enriched media with inorganic phosphate and glucose in high concentration (2xYTPG) has shown improved protein production yield [28]. In addition, enhanced central catabolic pathway and tricarboxylic acid (TCA) cycle support energy regeneration during CFPS reaction [29,30]. Rapidly growing cells maintain an increased catabolic and anabolic protein synthesis balance with sufficient nutrition supplement, and intracellular mass, including ribosome and volume of a bacterial cell increase exponentially as well [31,32]. Collection of bacterial cells at the exponential growth is known to be critical for obtaining more active cell extracts. For example, Kwon and Jewett examined the increased activity of cell extract when the cells were harvested at the exponential phase by monitoring cell growth rate [22].

In this study, one liter of 2xYTPG culture media was used for *E. coli* K12 MG1655 ΔprfA ΔendA Δrne cell culture. Overnight cultured cells (20 mL in LB) were inoculated to the 2xYTPG main culture media. Initial OD_{600} was measured from the overnight cultured cell. The cells were collected over the course of the culture, early mid-exponential phase, mid-exponential phase, early stationary phase, and stationary phase, which are marked with the pink arrows in Figure 2a. The cell extract was prepared with optimized conditions (sonication energy and cell-buffer volume) described previously [22]. The CFPS reaction with each cell extract was examined to determine the best harvest time point for maximizing protein productivity. Notably, the cell extract harvested at the mid-exponential growth phase (7 h culture time) showed significantly increased protein productivity (32.8%) compared to that of the extract harvested in the stationary growth phase (Figure 2b). This result indicated that the performance of cell-extract imperatively depends on the contents of intracellular macromolecules, which change during cell growth. Determination of mid-exponential growth for the cell collection is the first step to obtain the highly active cell extract for the CFPS reaction.

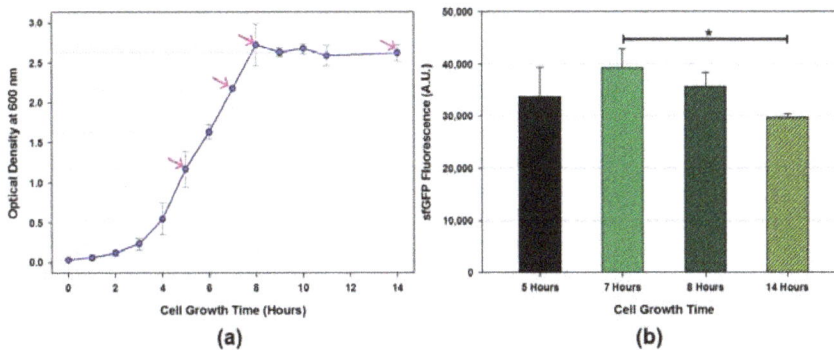

Figure 2. Cell extract performance upon the different cell harvest time. (**a**) The *E. coli* K12 MG1655 $\Delta prfA$ $\Delta endA$ Δrne growth curve. The cell collected at 5, 7, 8, 14 h from the beginning of cell culture. The pink arrows indicate each harvest time. (**b**) The intensity of the cell-free synthesized sfGFP fluorescence from each cell extract (1750 Joules of sonication energy input, 1000 μL of processing volume, and 60 min of runoff reaction time). Data are presented as the average ± standard deviation ($n = 3$); * $p < 0.05$ according to one-way analysis of variance (ANOVA).

4.2. Lysis

Kwon and Jewett developed a method to determine the proper level of sonication energy input per sample volume, which is required to disrupt *E. coli* cell wall without damaging the intercellular components. This cell extract showed the overall same protein production capacity compared with already established cell extract preparation methods [22].

In this study, the optimal lysis condition was determined with two independent variables: the processing volume for sonication and sonication energy input as described previously [22]. The aim of this study is to investigate the variations in the protein productivity of cell extracts, which is the result of the combinatorial change of the processing volume and sonication energy input. First, ten different sonication energy levels (50, 100, 250, 500, 750, 1000, 1250, 1500, 1750, and 2000 Joules) were applied to each of four different processing volumes (250, 500, 750, and 1000 μL) to determine the optimal cell lysis condition during sonication. The relative productivity of the cell extract was determined by the fluorescence intensity of the cell-free synthesized protein, sfGFP, divided by the fluorescence intensity from the cell extract showing the highest intensity at each processing volume. Then the relative intensity was represented as 0 to 100% for processing volume. The 3D mesh plot indicates the maximum protein productivity—color represents relative productivity—at each volume of cell extract tested against the sonication energy input resulting in the optimal conditions listed here: 100 J of sonication energy input for 250 μL, 500 to 1000 J for 500 μL, 1250 J for 750 μL, and 1750 J for 1000 μL (Figure 3a). Notably, 500 μL of processing volume showed a wide range of sonication energy input. Since the observed optimal sonication energy input at each processing volume ranged from 250 to 1000 μL fits the linear regression model in Figure 3c, the optimal sonication energy input for any processing volume ranged between 250 and 1000 μL can be predicted to obtain the best performing cell extract. The relative protein productivity is listed in Supplementary Table S1. The optimal sonication input for processing volume can be projected using the linear regression model as follows: $y = 2.18x - 400$ where x is the designed processing volume for sonication. The cell-free synthesized sfGFP from the cell extract variants of processing volume 750 μL were analyzed by Coomassie Blue staining after running the reaction samples on a 4–12% Bis-Tris acrylamide gel (Supplementary Figure S1). Next, all the variations in the protein productivity of cell extracts from the combined alterations of sonication energy input and processing volume were profiled. The cell extract lysed with the 1750 Joules of energy input and the 1000 μL of processing volume showed the overall highest protein productivity among the forty energy-volume combinations (Figure 3b). The relative productivity of cell extract was determined by the fluorescence intensity of the cell-free synthesized sfGFP divided by the sfGFP fluorescence intensity from the cell extract showing the highest intensity in all combined modifications. The relative intensity (combined) was represented as 0 to 100%. The result indicated that the higher volumes, such as 750 and 1000 µL of the processing volume tend to be more favorable to obtain more active cell extract when combined with the optimal sonication energy input (Supplementary Table S2). The linear relationship between sonication volume and energy input was plotted in Supplementary Figure S2. Lastly, we measured the total amount of protein in the cell extract. The total protein concentration in the cell extract increased by up to 54.78 mg/mL with the increase of sonication energy, and then showed a plateau in the range of 500–2000 Joules. This trend is clearly shown in the 500, 750, and 1000 µL of the processing volume (Figure 3d). Interestingly, the variation in the total protein of the cell extract was not parallel to the protein productivity of the cell extract (Supplementary Figure S3a–c). Although it is difficult to determine that the exact amount of required total protein in cell extract for the optimal protein production performance, approximately 45 to 50 mg/mL of total protein contents showed 60–100% of relative protein productivity (Supplementary Figure S3). The cell extract with low processing volume (250 µL) showed declining trends along with the increase of the sonication energy input (Figure 3d and Supplementary Figure S3d). Protein degradation by overheating is an inevitable drawback of the sonication process. To minimize possible heat degradation, we carried out sonication by 10 s lysis and 10 s cooling during the total input energy reached to our preset values in ice water as described previously [22]. Moreover, we assumed that the higher energy which is out of range of the optimal value is a lethal effect to the cell extract activity by disruption of all the macromolecule contents in the cell extract.

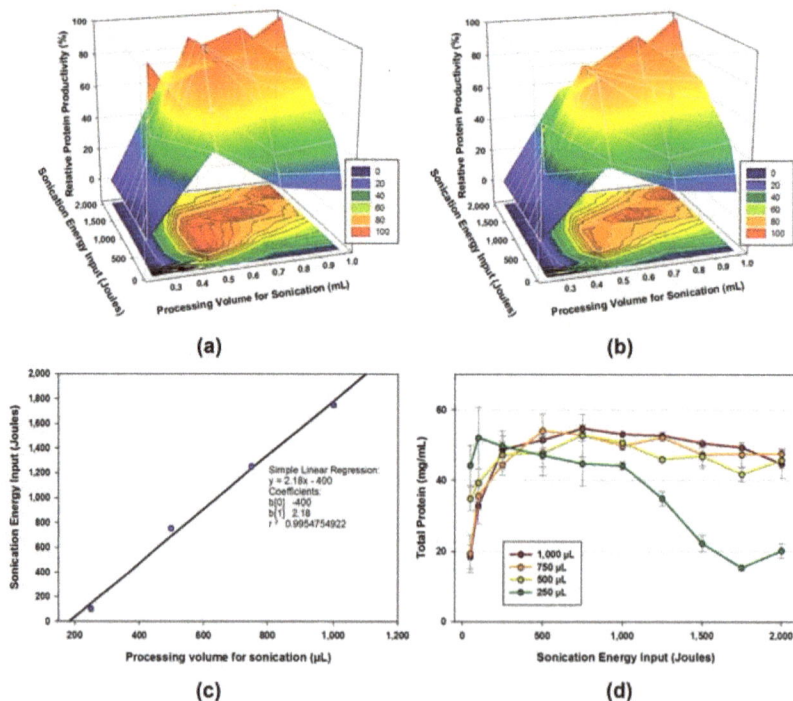

Figure 3. The effect of total energy input and cell suspension volume on the protein productivity of cell extract. (**a**) The relative protein productivity per each volume. The productivity was represented from 0 to 100% at each processing volume (250, 500, 750, and 1000 µL) by different energy input (from 50 J to 2000 J) to find the optimal sonication energy input for each processing volume. (**b**) The relative protein

productivity with all processing volume (250 to 1000 µL, combined) was represented from 0 to 100% by different energy input (from 50 J to 2000 J) to find the optimal sonication energy and volume for the highest protein productivity. (**c**) A simple linear regression model of processing volume and sonication energy input. The sonication energy input resulting in the highly active cell extract was selected for each processing volume (250, 500, 750, and 1000 µL) and plotted in a regression model. From the regression model, the optimal energy for each processing volume can be calculated as follows: (Sonication energy input (Joules)) = (The designed processing volume for sonication (µL) + 400) 2.18^{-1}. (**d**) The changes in the total protein amount of the cell extract in different sonication energy input. Data are presented as the average ± standard deviation ($n = 2$).

4.3. Post-Lysis Procedure

Runoff reaction and dialysis were considered as cell extract clarification steps in traditional cell extract preparation. However, commercially available *E. coli* strain BL21 showed robust cell extract performance without these additional processes after lysis [19]. Kwon and Jewett further optimized a correlation between runoff reaction time and cell extract performance for *E. coli* strains BL21(DE3) star and a partially engineered C495 [22]. Here, we investigated the effect of runoff reaction and dialysis on the protein productivity of cell extract derived from genomically engineered *E. coli* strain.

The different runoff reaction was conducted after lysis and centrifugation. The CFPS reaction was carried out for 20 h at 30 °C with pJL1-sfGFP and PCR amplified linear sfGFP template. The data on protein productivity by different runoff time in Figure 4a,b shows that the runoff reaction time is essential for the cell extract from *E. coli* K12 MG1655 Δ*prfA* Δ*endA* Δ*rne*. Protein production performance is comparable to the results in 40–80 min of runoff incubation time reported previously [21,22]. The statistical data using the one-way ANOVA indicates that there were significant differences ($p < 0.001$) between the cell extract with and without runoff reactions. However, there were no significant differences between cell extract within the 20–80 runoff reaction times. In addition, we observed that total protein contents were preserved during runoff reaction (Figure 4b). Next, we studied the effect of runoff reaction on all different sonication condition (50–1500 Joules, 1000 µL volume processing). The overall result is well correlated to our previous findings (Figure 4c and Supplementary Figure S3a–d). Regardless of sonication optimization, the significant increase of protein productivity was detected after runoff reaction ($p < 0.001$). Lastly, the effect of runoff reaction time using different DNA template was investigated. We carried out the CFPS reaction for 4 h for both PCR amplified linear template and plasmid instead of applying CFPS reaction for 20 h due to the instability of linear template. In both groups, the cell extract which was prepared without runoff reaction showed significantly low productivity (* $p < 0.05$) compared to the extract with runoff reaction (Figure 4d). Interestingly, there was no significant difference between the extracts with 20–80 min runoff reaction for the model genomically engineered *E. coli* strain.

After the cell lysis, centrifugation, and runoff reaction, the remained byproduct can be cleared from the cell extract by dialysis. While the conventional cell extract preparation method included the dialysis step for final cell extract [18], more recently optimized cell extract preparation methods showed the same protein production performance without this step [19,22,33]. In this study, we applied the dialysis as the final optional step in the entire process. The cell extract prepared as followed: 1750 Joules of energy input, 1000 µL processing volume, and 60 min of runoff reaction. The dialysis was carried out with 2 mL of cell extract in a dialysis bag (10K MWCO) at 4 °C. Buffer A was replaced every 45 min for four times. Although total protein concentration is preserved after dialysis, protein productivity was decreased by 22% (Figure 5a,b).

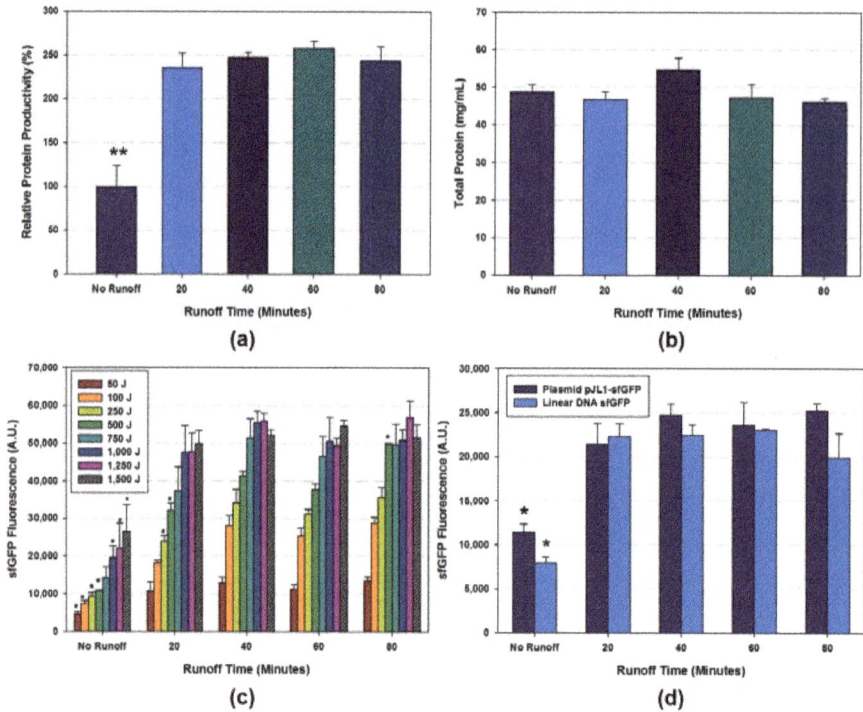

Figure 4. The effect of runoff reaction on protein productivity and total protein of the cell extract. (a) The relative protein productivity using the cell extract produced by different runoff reaction time ranged 0–80 min. Data depicted as the mean ± SD (*n* = 3); ** *p* < 0.001 according to one-way analysis of variance (ANOVA) (b) The total protein concentration in the cell extract treated with different runoff reaction time. For Figure 4a,b, the same cell lysate was used with the sonication condition of 1500 Joules of sonication input and 1000 μL of processing volume. (c) The effect of runoff at different lysis condition. All lysis procedure was fixed to 1000 μL of processing volume for sonication with different energy input ranged from 50 to 1500 Joules. Data depicted as the mean ± SD (*n* = 3), * *p* < 0.001 according to one-way analysis of variance (ANOVA). (d) CFPS reaction with runoff time variants on plasmid and PCR amplified linear template. Data depicted as the mean ± SD (*n* = 3), * *p* < 0.05 according to one-way analysis of variance (ANOVA).

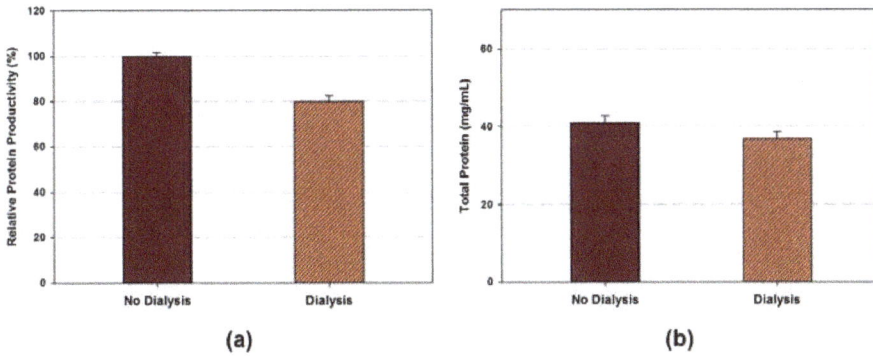

Figure 5. (a) The relative protein productivity, (b) total protein (mg/mL) of cell extract in the condition of dialysis. Data are presented as the average ± standard deviation (*n* = 2).

5. Conclusions

Over the past decade, the cell-free system has been revitalized as an essential toolkit for synthetic biology and biotechnology research. Moreover, with its unique non-living feature, the cell-free system provides a new insight to understand the cellular processes outside the shell. Preparation of a highly active cell extract is the first step to build this versatile CFPS system. In this study, we discussed a detailed cell extract preparation protocol in three major stages encompassing cell culture, sonication, and optional post-lysis steps for genomically engineered *E. coli* K12 MG1655 Δ*prfA* Δ*endA* Δ*rne* as a model strain. We expect that this protocol will provide not only a practical guideline but also a foundation for the entire CFPS system.

Supplementary Materials: The following are available online at http://www.mdpi.com/2409-9279/2/3/68/s1, Table S1: The relative protein productivity from the different sonication energy input at each processing volume, Table S2: The relative protein productivity with all processing volume (250–1000 μL, combined), Figure S1: The SDS-PAGE of the cell-free reaction mixture after reaction at 37 °C for 24 h, Figure S2: The linear relationship between sonication volume and energy input, Figure S3: The total protein (mg/mL) and relative sfGFP productivity (%) of cell extract variants of different sonication energy input ranged from 50 J to 2000 J.

Author Contributions: Experiment design and conceptualization, J.K., C.E.C., and Y.-C.K.; data acquisition and analysis, J.K., C.E.C., and S.R.P.; writing—original draft preparation, J.K.; writing—review and editing, J.K., C.E.C., S.R.P., and Y.-C.K.; supervision, Y.-C.K.

Funding: This research was supported by National Science Foundation EPSCoR OIA (Award No. 1632854).

Acknowledgments: We thank Michael C. Jewett (Northwestern University) for providing plasmid pJL1-sfGFP and *E. coli* strain K12 MG1655 Δ*prfA* Δ*endA* Δ*rne*.

Conflicts of Interest: The authors declare no conflict of interest.

References

1. Roberts, R.B. *Microsomal Particles and Protein Synthesis*; Pergamon Press: Oxford, UK, 1958.
2. Gregorio, N.E.; Levine, M.Z.; Oza, J.P. A user's guide to cell-free protein synthesis. *Methods Protoc.* **2019**, *2*, 24. [CrossRef]
3. Dopp, B.J.L.; Tamiev, D.D.; Reuel, N.F. Cell-free supplement mixtures: Elucidating the history and biochemical utility of additives used to support in vitro protein synthesis in *E. coli* extract. *Biotechnol. Adv.* **2019**, *37*, 246–258. [CrossRef] [PubMed]
4. Carlson, E.D.; Gan, R.; Hodgman, C.E.; Jewett, M.C. Cell-free protein synthesis: Applications come of age. *Biotechnol. Adv.* **2012**, *30*, 1185–1194. [CrossRef] [PubMed]
5. Levine, M.Z.; Gregorio, N.E.; Jewett, M.C.; Watts, K.R.; Oza, J.P. Escherichia coli-based cell-free protein synthesis: Protocols for a robust, flexible, and accessible platform technology. *JoVE* **2019**, *144*, e58882. [CrossRef] [PubMed]

6. Karim, A.S.; Heggestad, J.T.; Crowe, S.A.; Jewett, M.C. Controlling cell-free metabolism through physiochemical perturbations. *Metab. Eng.* **2018**, *45*, 86–94. [CrossRef] [PubMed]

7. Lim, H.J.; Kim, D.M. Cell-free metabolic engineering: Recent developments and future prospects. *Methods Protoc.* **2019**, *2*, 33. [CrossRef] [PubMed]

8. Casteleijn, M.G.; Urtti, A.; Sarkhel, S. Expression without boundaries: Cell-free protein synthesis in pharmaceutical research. *Int. J. Pharm.* **2013**, *440*, 39–47. [CrossRef] [PubMed]

9. Lee, S.H.; Kwon, Y.C.; Kim, D.M.; Park, C.B. Cytochrome P450-catalyzed O-dealkylation coupled with photochemical NADPH regeneration. *Biotechnol. Bioeng.* **2013**, *110*, 383–390. [CrossRef] [PubMed]

10. Ahn, J.H.; Chu, H.S.; Kim, T.W.; Oh, I.S.; Choi, C.Y.; Hahn, G.H.; Park, C.G.; Kim, D.M. Cell-free synthesis of recombinant proteins from PCR-amplified genes at a comparable productivity to that of plasmid-based reactions. *Biochem. Biophys. Res. Commun.* **2005**, *338*, 1346–1352. [CrossRef] [PubMed]

11. Kwon, Y.C.; Lee, K.H.; Kim, H.C.; Han, K.; Seo, J.H.; Kim, B.G.; Kim, D.M. Cloning-independent expression and analysis of omega-transaminases by use of a cell-free protein synthesis system. *Appl. Environ. Microbiol.* **2010**, *76*, 6295–6298. [CrossRef] [PubMed]

12. Sun, Z.Z.; Yeung, E.; Hayes, C.A.; Noireaux, V.; Murray, R.M. Linear DNA for rapid prototyping of synthetic biological circuits in an Escherichia coli based TX-TL cell-free system. *ACS Synth. Biol.* **2014**, *3*, 387–397. [CrossRef] [PubMed]

13. Son, J.-M.; Ahn, J.-H.; Hwang, M.-Y.; Park, C.-G.; Choi, C.-Y.; Kim, D.-M. Enhancing the efficiency of cell-free protein synthesis through the polymerase-chain-reaction-based addition of a translation enhancer sequence and the in situ removal of the extra amino acid residues. *Anal. Biochem.* **2006**, *351*, 187–192. [CrossRef] [PubMed]

14. Hong, S.H.; Kwon, Y.C.; Jewett, M.C. Non-standard amino acid incorporation into proteins using *Escherichia coli* cell-free protein synthesis. *Front. Chem.* **2014**, *2*, 34. [CrossRef] [PubMed]

15. Martin, R.W.; Des Soye, B.J.; Kwon, Y.C.; Kay, J.; Davis, R.G.; Thomas, P.M.; Majewska, N.I.; Chen, C.X.; Marcum, R.D.; Weiss, M.G.; et al. Cell-free protein synthesis from genomically recoded bacteria enables multisite incorporation of noncanonical amino acids. *Nat. Commun.* **2018**, *9*, 1203. [CrossRef] [PubMed]

16. Pardee, K.; Green, A.A.; Ferrante, T.; Cameron, D.E.; DaleyKeyser, A.; Yin, P.; Collins, J.J. Paper-based synthetic gene networks. *Cell* **2014**, *159*, 940–954. [CrossRef] [PubMed]

17. Kigawa, T.; Yabuki, T.; Matsuda, N.; Matsuda, T.; Nakajima, R.; Tanaka, A.; Yokoyama, S. Preparation of *Escherichia* coli cell extract for highly productive cell-free protein expression. *J. Struct. Funct. Genom.* **2004**, *5*, 63–68. [CrossRef] [PubMed]

18. Zubay, G. In vitro synthesis of protein in microbial systems. *Annu. Rev. Genet.* **1973**, *7*, 267–287. [CrossRef] [PubMed]

19. Kim, T.-W.; Keum, J.-W.; Oh, I.-S.; Choi, C.-Y.; Park, C.-G.; Kim, D.-M. Simple procedures for the construction of a robust and cost-effective cell-free protein synthesis system. *J. Biotechnol.* **2006**, *126*, 554–561. [CrossRef]

20. Yang, W.C.; Patel, K.G.; Wong, H.E.; Swartz, J.R. Simplifying and streamlining Escherichia coli-based cell-free protein synthesis. *Biotechnol. Prog.* **2012**, *28*, 413–420. [CrossRef]

21. Liu, D.V.; Zawada, J.F.; Swartz, J.R. Streamlining Escherichia coli S30 extract preparation for economical cell-free protein synthesis. *Biotechnol. Prog.* **2005**, *21*, 460–465. [CrossRef]

22. Kwon, Y.-C.; Jewett, M.C. High-throughput preparation methods of crude extract for robust cell-free protein synthesis. *Sci. Rep.* **2015**, *5*, 8663. [CrossRef] [PubMed]

23. Pratt, J.M. *Coupled Transcription-Translation in Prokaryotic Cell-Free Systems*; IRL Press: Oxford, UK, 1984; pp. 179–209.

24. Goerke, A.R.; Swartz, J.R. Development of cell-free protein synthesis platforms for disulfide bonded proteins. *Biotechnol. Bioeng.* **2008**, *99*, 351–367. [CrossRef] [PubMed]

25. Salehi, A.S.M.; Smith, M.T.; Bennett, A.M.; Williams, J.B.; Pitt, W.G.; Bundy, B.C. Cell-free protein synthesis of a cytotoxic cancer therapeutic: Onconase production and a just-add-water cell-free system. *Biotechnol. J.* **2016**, *11*, 274–281. [CrossRef] [PubMed]

26. Krinsky, N.; Kaduri, M.; Shainsky-Roitman, J.; Goldfeder, M.; Ivanir, E.; Benhar, I.; Shoham, Y.; Schroeder, A. A simple and rapid method for preparing a cell-free bacterial lysate for protein synthesis. *PLoS ONE* **2016**, *11*, e0165137. [CrossRef] [PubMed]

27. Lee, K.H.; Kim, D.M. Recent advances in development of cell-free protein synthesis systems for fast and efficient production of recombinant proteins. *FEMS Microbiol. Lett.* **2018**, *365*, fny174. [CrossRef] [PubMed]

28. Kim, R.G.; Choi, C.Y. Expression-independent consumption of substrates in cell-free expression system from *Escherichia coli*. *J. Biotechnol.* **2000**, *84*, 27–32. [CrossRef]

29. Jewett, M.C.; Swartz, J.R. Substrate replenishment extends protein synthesis with an in vitro translation system designed to mimic the cytoplasm. *Biotechnol. Bioeng.* **2004**, *87*, 465–471. [CrossRef] [PubMed]

30. Jewett, M.C.; Calhoun, K.A.; Voloshin, A.; Wuu, J.J.; Swartz, J.R. An integrated cell-free metabolic platform for protein production and synthetic biology. *Mol. Syst. Biol.* **2008**, *4*, 220. [CrossRef] [PubMed]

31. Bosdriesz, E.; Molenaar, D.; Teusink, B.; Bruggeman, F.J. How fast-growing bacteria robustly tune their ribosome concentration to approximate growth-rate maximization. *FEBS J.* **2015**, *282*, 2029–2044. [CrossRef]

32. Scott, M.; Klumpp, S.; Mateescu, E.M.; Hwa, T. Emergence of robust growth laws from optimal regulation of ribosome synthesis. *Mol. Syst. Biol.* **2014**, *10*, 747. [CrossRef]

33. Shrestha, P.; Holland, T.M.; Bundy, B.C. Streamlined extract preparation for Escherichia coli-based cell-free protein synthesis by sonication or bead vortex mixing. *Biotechniques* **2012**, *53*, 163–174. [CrossRef] [PubMed]

methods and protocols

MDPI

Benchmark

Optimizing Cell-Free Protein Synthesis for Increased Yield and Activity of Colicins

Xing Jin [1], Weston Kightlinger [2] and Seok Hoon Hong [1,*]

1   Department of Chemical and Biological Engineering, Illinois Institute of Technology, Chicago, IL 60616, USA;
     xjin14@hawk.iit.edu
2   Department of Chemical and Biological Engineering, Northwestern University, Evanston, IL 60208, USA;
     westonkightlinger@u.northwestern.edu
*   Correspondence: shong26@iit.edu; Tel.: +1-312-567-8950

Received: 23 February 2019; Accepted: 3 April 2019; Published: 11 April 2019

Abstract: Colicins are antimicrobial proteins produced by *Escherichia coli* that hold great promise as viable complements or alternatives to antibiotics. Cell-free protein synthesis (CFPS) is a useful production platform for toxic proteins because it eliminates the need to maintain cell viability, a common problem in cell-based production. Previously, we demonstrated that colicins produced by CFPS based on crude *Escherichia coli* lysates are effective in eradicating antibiotic-tolerant bacteria known as persisters. However, we also found that some colicins have poor solubility or low cell-killing activity. In this study, we improved the solubility of colicin M from 16% to nearly 100% by producing it in chaperone-enriched *E. coli* extracts, resulting in enhanced cell-killing activity. We also improved the cytotoxicity of colicin E3 by adding or co-expressing the E3 immunity protein during the CFPS reaction, suggesting that the E3 immunity protein enhances colicin E3 activity in addition to protecting the host strain. Finally, we confirmed our previous finding that active colicins can be rapidly synthesized by observing colicin E1 production over time in CFPS. Within three hours of CFPS incubation, colicin E1 reached its maximum production yield and maintained high cytotoxicity during longer incubations up to 20 h. Taken together, our findings indicate that colicin production can be easily optimized for improved solubility and activity using the CFPS platform.

Keywords: colicins; cell-free protein synthesis; antimicrobials; chaperones

1. Introduction

Multidrug-resistant bacteria can be difficult to treat and are a serious threat to society [1]. There is an immediate need for the development of new antimicrobial drugs to counteract the increase in drug-resistant pathogens and the weakness of current antibiotic discovery pipelines [2]. A subgroup of antimicrobial peptides/proteins known as bacteriocins are considered to be viable alternatives to antibiotics because they exhibit high cell-killing activity against clinically important pathogens (both in vivo and in vitro), low oral toxicity to the host, as well as both broad- and narrow-spectrum qualities. Furthermore, bacteriocins can be produced by probiotic bacteria and easily bioengineered via protein engineering [3].

Antimicrobial peptides are short (5~100 amino acids) and can be synthesized either chemically or biologically [4]; however, the complete chemical synthesis of high-molecular weight proteins is challenging and generally require biological production by expression of corresponding genes in a host strain or in vitro system [5]. Compared to cell-based protein production, cell-free protein synthesis (CFPS) provides several advantages for producing toxic proteins, as demonstrated in the cases of onconase (RNase) [6], pierisin-1b [7], cecropin P1 [8], and colicins [9], which are deleterious to the host cells when overproduced. Most importantly, CFPS platforms do not need to maintain host cell viability during protein production because the transcriptional and translational machineries have already been extracted from lysed cells [10]. In addition, CFPS platforms provide an open reaction environment that can be easily controlled and optimized [11,12]. Our group recently demonstrated that colicins, bacteriocins produced by *Escherichia coli*, can be produced using an *E. coli*-based CFPS system [9]. Colicins kill non-host *E. coli* cells [13] by inhibiting cell wall synthesis (e.g., colicin M) [14], forming pores in inner membrane (e.g., colicin E1 [15] and Ia [16]), and degrading DNA (e.g., colicin E2) [17] or RNA (e.g., colicin E3) [18]. We reported that colicins E1 and E2 are very effective in killing antibiotic-tolerant persister cells [9]. However, some colicins such as colicin M exhibited low solubility and poor cell-killing activity when produced in CFPS. Therefore, further improvement of colicin production and bioactivity is required for optimal cell-free colicin production.

When the proteins are not folded properly, they become insoluble and form inclusion bodies in the cell [19]. Because proper three-dimensional protein folding is critical to achieve full protein function, these insoluble proteins are generally inactive [20]. Several strategies are available to improve solubility and thereby enhance protein folding including reducing protein synthesis rate, changing the growth medium, co-expressing molecular chaperones or foldases, and adding fusion partners to the target protein [21]. Similar approaches have been applied to produce 'difficult-to-express' proteins in vitro by harnessing the open and flexible nature of the CFPS platform [10]. Molecular chaperones prevent protein aggregation and promote protein folding via ingenious mechanisms [22]. The exogenous addition of molecular chaperone proteins has successfully facilitated the solubility of hundreds of proteins in cell-free translation system [23,24]. Common examples of molecular chaperones that can be used to prevent protein aggregation and misfolding include the GroES/EL and DnaK/DnaJ/GrpE chaperone systems [25]. For instance, cell extracts enriched with GroES/EL chaperones have been used to increase the yield of functional antibody fragments [26], and the solubility of the human erythropoietin was dramatically enhanced by cell extracts enriched with DnaK/DnaJ/GrpE chaperones [27]. Based on these previous works, we reasoned that CFPS could be optimized to improve colicin production yields and activity.

In this study, we investigated whether the production of colicins exhibiting low solubility, low yield, or low activity can be improved by optimizing CFPS lysates and reaction conditions. Here we report that the solubility of colicin M can be improved from 16% to nearly 100% by using CFPS lysates enriched with chaperones and that the cell-killing activity of colicin E3 can be increased by five orders of magnitude by co-expressing its immunity protein in the CFPS reaction. In addition, we measured the kinetics of colicin E1 synthesis and observed rapid formation of active colicin in CFPS reactions. This work provides new strategies to produce high titers of active colicins in CFPS and finds that CFPS is an excellent biological platform for the production and optimization of toxic proteins.

2. Materials and Methods

2.1. Bacterial Strains and Plasmids

The bacterial strains and plasmids used in this study are found in Table 1. We used BL21 Star (DE3) to prepare crude extracts because the strain carries a genomic copy of T7 RNA polymerase and a mutation in the RNase E gene (*rne131*) which facilitate the transcription and prevent the degradation of messenger RNAs, respectively [28], during the CFPS reaction [29]. We used the *E. coli* K361 strain for testing colicin cell-killing activity [30]. Plasmids containing molecular chaperone genes were purchased (Takara Bio, Shiga, Japan). These plasmids carry an origin of replication derived from pACYC and a gene providing chloramphenicol resistance. The chaperone genes are controlled by the *araB* or *Pzt-1* promoters which are induced by L-arabinose or tetracycline, respectively (Table 1). Detailed plasmid information can be found in the product manual provided from Takara Bio. Streptomycin (100 µg/mL), kanamycin (50 µg/mL), ampicillin (100 µg/mL) and chloramphenicol (35 µg/mL) were added in the cell culture to maintain plasmids as necessary.

2.2. Crude Extract Preparation

E. coli crude extracts for CFPS were prepared from the BL21 Star (DE3) strain using a sonication method [31] as described previously [9]. Chaperone-enriched cell extracts were prepared with similar methods after induction of chaperone protein production during *E. coli* cultivation. An overnight culture of BL21 Star (DE3) containing chaperone plasmids was diluted 1,000 times in 1.0 L of 2xYTPG medium (16 g/L tryptone, 10 g/L yeast extract, 5 g/L NaCl, 7 g/L K_2HPO_4, 3 g/L KH_2PO_4, and 18 g/L glucose; pH 7.2 adjusted with KOH) with appropriate antibiotics in a 2.5 L Tunair flask. Cells were grown to a turbidity at 600 nm (OD_{600nm}) of 0.5 at 37 °C at 220 rpm. The GroES/EL chaperone system from pGro7 was induced by adding 0.5 mg/mL L-arabinose. Similarly, DnaK/DnaJ/GrpE system from pKJE7 was induced by adding 0.5 mg/mL L-arabinose. For cell cultures containing both chaperone systems from pG-KJE8, GroES/EL was induced by adding 1 ng/mL tetracycline and DnaK/DnaJ/GrpE was induced by adding 0.5 mg/mL L-arabinose. After inducing chaperones, the cells were further grown to an OD_{600nm} of 3.0. Then, cells were harvested by centrifuging at 5,000× *g* at 4 °C for 15 min, washed twice to remove all the medium with cold S30 buffer (10 mM tris-acetate pH 8.2, 14 mM magnesium acetate, 60 mM potassium acetate, 1 mM dithiothreitol), and stored at −80 °C. Thawed cells were mixed with S30 buffer (1 mL buffer per 1 g of cells) and lysed on ice using a Q125 sonicator (Qsonica, Newtown, CT, USA) using 50% amplitude and three cycles of 45 s pulses at 60 s intervals. Insoluble components including cell debris were removed by two centrifugation steps at 14,000× *g* at 4 °C for 10 min. The crude extracts were filtered by a 0.2 µm sterile syringe filter (Corning, Corning, NY, USA) to remove unlysed cells completely. The total protein concentration of the extracts was approximately 50 mg/mL, as assessed by Quick-Start Bradford protein assay kit (Bio-Rad, Hercules, CA, USA). The crude extracts were stored at −80 °C until needed.

Table 1. Strains and plasmids used in this study. Str^R, Km^R, Am^R, and Cm^R are streptomycin, kanamycin, ampicillin, and chloramphenicol resistance, respectively.

Strains and Plasmids	Genotype/Relevant Characteristics	Source
Strains		
E. coli K361	Wild type W3110 strain with Str^R	[30]
E. coli BL21 Star (DE3)	F⁻ *ompT*, *hsdS*$_B$ (r$_B$⁻m$_B$⁻), *gal*, *dcm*, *rne131* (DE3)	Invitrogen
E. coli TG1	Strain containing colicin plasmid	[32]
Plasmids		
pJL1-sfGFP	Km^R, P_{T7}::*sfGFP*, C-terminal Strep-tag	[33]
pJL1-*cma*	Km^R, P_{T7}::*cma* encoding colicin M	[9]
pJL1-E3 *imm*	Km^R, P_{T7}::*E3imm* encoding E3 immunity	This study
pKSJ331	Am^R, ColE1 operon	[32]
pKSJ167	Am^R, ColE3 operon	[34]
pGro7	Cm^R, P_{araB}::*groES-groEL* encoding GroES/EL	Takara Bio
pKJE7	Cm^R, P_{araB}::*dnaK-dnaJ-grpE* encoding DnaK/DnaJ/GrpE	Takara Bio
pG-KJE8	Cm^R, P_{araB}::*dnaK-dnaJ-grpE* encoding DnaK/DnaJ/GrpE, P_{Pzt-1}::*groES-groEL* encoding GroES/EL	Takara Bio

2.3. Preparing Linear DNA Template for Cell-Free Colicin Production

Colicin genes were amplified by polymerase chain reaction (PCR) using primers found in Table S1. The *cma* gene encoding colicin M was amplified from the pJL1-*cma* plasmid using GAcol-F and GAcol-R, *ceaC* encoding colicin E3 from the pKSJ167 plasmid using ColE3-F and ColE3-R, and *cea* encoding colicin E1 from the pKSJ331 plasmid using ColE1-F and ColE1-R. PCR was performed using Phusion High-Fidelity DNA polymerase (New England Biolabs, Ipswich, MA, USA) at 98 °C for 30 s, with 30 cycles of denaturing at 98 °C for 10 s, annealing at 55 or 60 °C for 30 s, and extending at 72 °C for 2 min 30 s, and a final extension at 72 °C for 10 min. Three rounds of PCR were performed to insert T7 promoter and T7 terminator sequences for genes of colicin M, E1, and E3. Phosphorothioated T7Mega-F and T7Mega-R primers were used to protect T7 promoter and terminator sequences from nuclease degradation. The first PCR amplified the colicin genes, the second PCR added the T7 promoter sequence, and the third PCR added the T7 terminator sequence. DNA sequences of all colicin genes are found in Table S2. The PCR products were purified using E.Z.N.A. Cycle Pure kit (Omega Bio-Tek, Norcross, GA, USA) before addition to CFPS reactions.

2.4. Preparing E3 Immunity Protein

The E3 immunity gene *imm* was cloned into the pJL1 plasmid. First, the gene was amplified from the pKSJ167 plasmid using the primers E3Imm-F and E3Imm-R which installed a C-terminal Strep-tag (WSHPQFEK). The PCR fragment was double digested by NdeI and SalI restriction enzymes and ligated into pJL1-backbone at the same restriction sites. BL21 Star (DE3) cells were transformed with the ligated plasmid. An overnight culture of *E. coli* BL21 Star (DE3) harboring the pJL1-E3 *imm* plasmid was diluted 100 times in 250 mL of LB containing 50 μg/mL of kanamycin in a 1 L flask and incubated at 37 °C at 220 rpm until $OD_{600nm} \sim 0.5$. Then, E3 immunity protein production was induced by adding 1 mM of isopropyl β-D-1-thiogalactopyranoside at 37 °C at 220 rpm for an additional 5 h. E3 immunity protein was purified by a Strep-Tactin column (IBA, Göttingen, Germany), and the concentration of the E3 immunity protein was measured as 520 ± 30 μg/mL.

2.5. CFPS Reaction

CFPS reactions were performed to synthesize colicins according to established protocols [9]. Chaperone-enriched or standard BL21 Star (DE3) extracts were used, and RNase Inhibitor (New England Biolabs) or purified E3 immunity protein were added as necessary. The CFPS samples (15 μL in a 1.5 mL microcentrifuge tube) were incubated for up to 20 h at 30 °C or room temperature (~25 °C).

2.6. Quantifying Colicins Using Radioactive ^{14}C-Leu Assay

Total and soluble protein yields were measured by determining radioactive ^{14}C-Leu incorporation [35]. Briefly, 10 μM ^{14}C-leucine (Perkin-Elmer, Waltham, MA, USA) was added into triplicate CFPS reactions. After incubating at 30 °C or room temperature for specified periods of time, soluble fractions were separated by centrifuging at 12,000× *g* for 15 min at 4 °C. CFPS reactions in the time-course study of colicin E1 synthesis were quenched at indicated times using 833 μg/mL kanamycin and flash freezing at −80 °C. Proteins were precipitated, washed with 5% trichloroacetic acid three times and then 100% ethanol, and quantified using a liquid scintillation counting. To eliminate background radioactivity and protein synthesis, scintillation counts from no plasmid control were subtracted from the colicin samples. For colicin M, total or soluble fractions of each reaction containing ^{14}C-leucine were visualized by running an SDS-PAGE gel, exposing the gel Storage Phosphor Screen, and acquiring an autoradiogram using a Typhoon FLA700 imager as described previously [9].

2.7. Cell Viability Test

Cell viability assays were performed as described previously [9]. An overnight culture of the K361 indicator cells was regrown in fresh LB medium at 220 rpm at 37 °C until an OD_{600nm} of approximately 0.7–0.9. Cells were harvested, adjusted to an OD_{600nm} of 0.1 (equivalent to 5.0×10^7 CFU/mL) or 1.0 (equivalent to 5.0×10^8 CFU/mL) with LB medium, and then incubated with CFPS reaction products for 1 h with shaking at 37 °C. Cell viability was quantified by counting colony forming units. To generate the activity curve of colicin E3, we treated cells at high initial population (5.0×10^8 CFU/mL) with varying concentrations of colicin E3 for 1 h. Effective multiplicity (m), a measure of colicin cytotoxicity, was calculated by the equation: m = −ln(S/S$_0$), where S indicates the surviving cell population with colicin treatment, and S$_0$ is the untreated control cell population [36].

2.8. Statistical Analysis

Two-tailed *t*-tests between colicin samples and no colicin controls were performed [9]. Statistical significance is indicated in figures with * ($p < 0.01$), ** ($p < 0.001$), and *** ($p < 0.0001$).

3. Results and Discussion

3.1. Enrichment of Cell Extracts with Chaperones Does Not Significantly Affect CFPS Productivity

While our previous study showed that colicins E1, E2 and Ia were nearly completely soluble when produced in CFPS, we found that the majority of colicin M produced was insoluble and that only soluble colicin M (~5%) was active in killing K361 indicator cells [9]. Based on previous reports that cell extracts enriched with chaperones and disulfide bond isomerases can enhance the production of functional antibodies in CFPS [26], we sought to apply a similar strategy to improve the solubility of colicin M. We overexpressed three sets of molecular chaperones (GroES/EL (Gro), DnaK/DnaJ/GrpE (KJE), and both GroES/EL and DnaK/DnaJ/GrpE (Gro-KJE)) in BL21 Star (DE3) *E. coli* cultures and then harvested and prepared extracts from these chassis strains using a sonication method [31] followed by the removal of unlysed cells by syringe filtration [9]. First, we examined if the overexpression of chaperones in CFPS extract chassis strains affected their overall in vitro protein production capacity by producing superfolder green fluorescent protein (sfGFP) (Figure S1). sfGFP is known as a highly soluble protein with rapid folding kinetics [37] and has been widely used as an indicator to examine the CFPS capacity of cell extracts [31]. The sfGFP production levels using the chaperone-enriched cell extracts were similar (within ± 20% difference) to the sfGFP level using the standard cell extract (Star) prepared from BL21 Star (DE3) without chaperone overexpression. As expected, we obtained almost 100% soluble sfGFP for all extracts tested. These results show that chaperones in the extracts do not strongly affect transcription and translation during CFPS reactions.

3.2. Solubility of Colicin M is Increased in the Presence of Chaperones in CFPS

Next, we produced colicin M by CFPS using the three chaperone-enriched extracts characterized above to determine if the chaperones could enhance solubility of colicin M. We incubated the cell-free reactions for 20 h at 30 °C and then quantified total and soluble colicin M production via radioactive ^{14}C-Leu incorporation. Total colicin M yield was approximately 300 ng/μL in all extracts except the GroES/EL-containing extract which yielded approximately 500 ng/μL (Figure 1A). The total colicin M yield in CFPS was comparable with other colicins (E1, E2, and Ia) previously reported [9]. Without chaperone-enriched cell extracts, less than 20% of colicin M was soluble. However, the Gro chaperone system increased the solubility to close to 30%, and the KJE chaperone system improved the solubility to more than 80%. When both chaperone systems were present in the cell extract (Gro-KJE), colicin M was completely soluble (Table 2, Figure 1A). The improvement of colicin M solubility by chaperones present in the CFPS reaction was confirmed by an SDS-PAGE gel autoradiogram (Figure 1B). Our results reveal that GroES/EL and DnaK/DnaJ/GrpE chaperones overexpressed in CFPS chassis strains can assist in protein folding in vitro, thereby enhancing the colicin M solubility and the titers of soluble colicin M which can be produced during cell-free reactions. Notably, the presence of both chaperone systems results in a synergistic effect leading to the production of colicin M that is 100% soluble.

A)

B)

C)

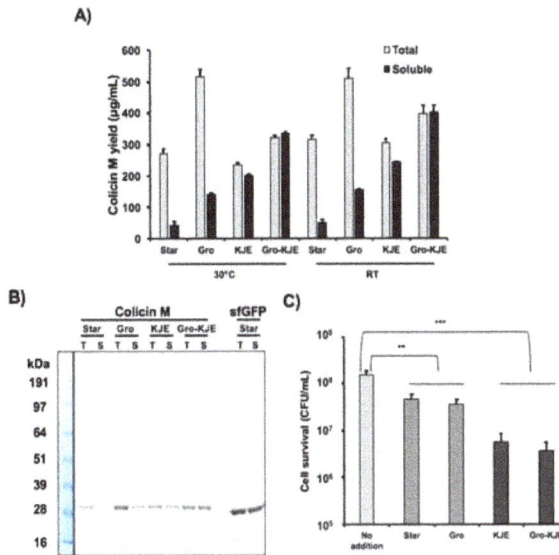

Figure 1. Improvement of colicin M production in cell-free protein synthesis (CFPS). (**A**) Total and soluble protein yield for cell-free produced colicin M with molecular chaperone-enriched extracts quantified by ^{14}C-Leu scintillation counting at 30 °C and room temperature (RT). Cell extract without chaperone production was prepared from BL21 Star (DE3) strain (Star). Chaperone-enriched extracts were prepared from BL21 Star (DE3) cultures overexpressing GroES/EL (Gro), DnaK/DnaJ/GrpE (KJE), and both GroES/EL and DnaK/DnaJ/GrpE (Gro-KJE) in BL21 Star (DE3). Error bars indicate standard deviations from three independent CFPS reactions. (**B**) Radioactive ^{14}C-Leu autoradiogram gel of total (T) and soluble (S) protein yield for colicin M and sfGFP produced during CFPS reactions with different cell extracts. (**C**) Viability of K361 indicator cells. K361 cells (initial cell density 5×10^7 CFU/mL) were treated with 750 ng/mL total concentration of colicin M produced by CFPS using different chaperone-enriched extracts and then incubated at 37 °C and 220 rpm for 1 h. Error bars indicate standard deviation from two independent cultures with three plating replicates each. ** and *** represent significant difference compared to no addition sample under the same media conditions with p-values of < 0.001 and <0.0001, respectively.

Table 2. Solubility of colicin M. Solubility percentages based on total and soluble yields of cell-free synthesized colicin M produced in various lysates and incubation temperatures and quantified by radioactive ^{14}C-Leu scintillation counting (yield data shown in Figure 1A). BL21 Star (DE3) extracts enriched with chaperones by overexpression of GroES/EL (Gro), DnaK/DnaJ/GrpE (KJE), and both GroES/EL and DnaK/DnaJ/GrpE (Gro-KJE) were used. BL21 Star (DE3) extract without chaperone overexpression (Star) was used as a control. RT indicates room temperature (~25 °C). Average ± standard deviations are shown.

Extracts	Solubility (%)	
	30 °C	RT
Star	16 ± 4	16 ± 3
Gro	27 ± 2	30 ± 2
KJE	86 ± 3	80 ± 3
Gro-KJE	104 ± 2	102 ± 9

Although we did not purify colicin M from the cell-free reaction that contains chaperones and other CFPS components, we expect that colicin M is still soluble and active when the chaperones are removed via purification. Previous work suggests that the molecular chaperones used in our cell-free reactions form protein-chaperone complexes to assists in folding, and that after folding, the protein is released from the complex to become a folded or native protein [38,39]. Another report demonstrated that a soluble single chain variable fragment (scFv) antibody can be successfully purified after co-expression with a molecular chaperone [40].

Decreasing incubation temperatures in cell cultures producing heterologous proteins is known to result in decreased protein aggregation [41]. This is likely due to slower protein production rates at lower temperatures which give newly translated proteins time to fold properly [21]. To examine whether or not lowering the temperature during protein synthesis could further influence colicin M production, we incubated cell-free reaction with and without molecular chaperones at room temperature (~25 °C) for 20 h. The incubation temperature did not affect colicin M solubility (Table 2), but the soluble colicin M production yields were improved 17 ± 5% at room temperature compared to the colicin M yield at 30 °C (Figure 1A). Similar low temperature effect in enhancing protein yield in CFPS was also reported in sfGFP synthesis [33] and ribosomal RNA synthesis and ribosome assembly [42]. Since the amount of soluble colicin M yield was increased at the lower incubation temperature, we produced colicin M via CFPS at room temperature for later study. Taken together, our data shows that cell extracts enriched with molecular chaperones enhance colicin M solubility and lowering the incubation temperature from 30 °C to room temperature increases colicin M yields.

3.3. Increased Colicin M Solubility Enhances Cell-Killing Activity

As the colicin M solubility was increased with the chaperone-enriched extracts, we investigated the cell-killlng activity of colicin M produced under these conditions. To prepare indicator cells for this experiment, a culture of *E. coli* K361 strain was incubated to reach exponential growth phase with a turbidity (OD_{600nm}) between approximately 0.7–0.9, centrifuged, and resuspended in fresh LB to adjust cell populations to approximately 5.0×10^7 CFU/mL. The indicator cells were exposed to a 750 ng/mL total concentration of cell-free produced colicin M for 1 h at 37 °C with shaking at 220 rpm (Figure 1C). We observed that colicin M produced by chaperone-enriched cell extracts (KJE and Gro-KJE extracts) lowered surviving cells up to 12-fold compared to the colicin M produced by the Star extract. The levels of cell survival observed in cultures treated with colicin M produced by KJE extract was similar to the cultures treated with colicin M produced by Gro-KJE extract (Figure 1C). Because the KJE extract already increased the colicin M solubility to over 80%, the solubility improvement to 100% achieved by using Gro-KJE extract (Table 2) might not result in significant differences in the apparent cell-killing activity of colicin M. We previously reported that only soluble colicin M exhibited cell-killlng activity [9], likely due to the improper folding and therefore a lack of activity of the insoluble proteins [19]. These results suggest that the increased solubility of colicin M achieved by production in lysates enriched with molecular chaperones (KJE and Gro-KJE extracts) enhances the overall cell-killing activity of colicin M.

3.4. Co-Expression of Colicin E3 and Its Immunity Protein Enhances E3 Activity

We previously reported that the production of colicin E2 (which has DNase activity) in CFPS does not require the addition or co-expression of its immunity protein. In this study, we were interested in producing another colicin (E3) that is known to have a specific RNase activity [18] which degrades the 16S ribosomal RNA [43]. We originally hypothesized that because of this RNase activity, colicin E3 would be difficult to produce in cell-free as it might digest the ribosomes which are required for its synthesis and that blocking this RNase activity during its CFPS production would increase overall yields. We did, in fact, observe low production yield of colicin E3 (~60 ng/μL) (Figure S2) which was 5-fold lower than that of colicin E1 and E2 (~300 ng/μL) in CFPS [9]. However, neither the addition of RNase inhibitor [44] nor the addition of exogenously produced E3 immunity protein (both of which would presumably block the RNase activity of colicin E3) were effective in increasing our overall yields (Figure S2). We also attempted to co-express the E3 immunity gene together with the colicin E3 gene in an operon during CFPS, but this co-expression also had no effect on the protein production yield (Figure S2). Despite the fact that the addition of RNase inhibitor and E3 immunity protein did not improve production yields, we decided to test the cytotoxicity of colicin E3 produced in CFPS with exogenous addition of RNase inhibitor or its immunity protein or co-expression of E3 and immunity protein (Figure 2A). The cell-killng activity of colicin E3 synthesized without the immunity protein or supplemented with RNase inhibitor present during the CFPS reaction possessed very little cell-killing activity. However, we were surprised to find that the exogenous addition of the E3 immunity protein into the CFPS reaction increased the E3 activity 1000-fold and that the co-expression of colicin E3 and immunity protein during the cell-free reaction increased the E3 activity 10^5-fold with a multiplicity of 13.5, killing nearly all indicator cells (Figure 2A).

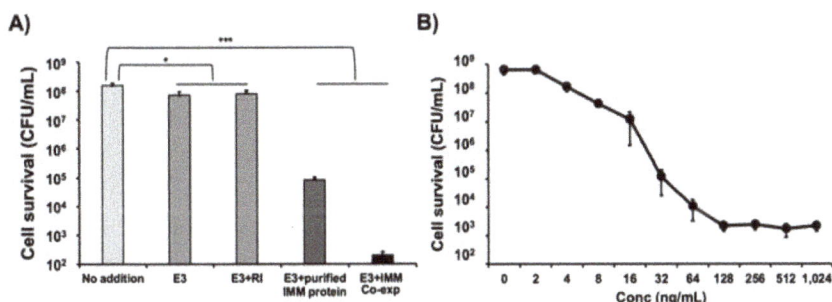

Figure 2. Improvement of colicin E3 activity in CFPS. (**A**) Viability of K361 indicator cells (initial cell density 5×10^7 CFU/mL) upon treatment with 250 ng/mL of cell-free produced colicin E3 with 2.7 U/μL RNase inhibitor, 16.7 ng/μL immunity protein, and co-expression (Co-exp) of immunity protein at 37 °C 220 rpm for 1 h. (**B**) Effect of increasing concentrations of cell-free produced colicin E3 with co-expression of immunity protein on K361 cells (initial cell density 5×10^8 CFU/mL) at 37 °C 220 rpm for 1 h incubation. Error bars indicate standard deviation from two independent cultures with three plating replicates each. * and *** represent significant differences compared to no addition sample under the same media conditions with *p*-values of < 0.01 and <0.0001, respectively.

Our results suggest that the immunity protein is necessary for colicin E3 to maintain its activity. While initially unexpected, this finding makes sense in the context of a structural study of colicin E3 [45] which found that the colicin E3 protein alone contains a disordered cytotoxicity domain that is restored to its native structure when complexed with the E3 immunity protein. We therefore conclude that colicin E3 produced alone in CFPS has very low activity due to a disordered cytotoxicity domain and that this cytotoxicity domain folds into its native structure and becomes active when the colicin E3 immunity protein is added to the reaction. Furthermore, we observed that co-expression of the immunity protein provided much more highly active colicin E3 compared to the addition of purified immunity protein. This difference suggests that there may be a close interaction between colicin E3 and the immunity protein that is required during their folding to obtain full activity from the complex. Using the co-expression approach, we produced fully active colicin E3 and assessed the cell-killing activity by varying concentrations of E3. Maximum cell-killng was achieved at concentrations above 128 ng/mL of E3 and the immunity complex with a multiplicity around 13 (Figure 2B) which is comparable to the activity of cell-free synthesized colicins E1 and E2 [9] and consistent with the previous report that the multiplicity of colicin E3 produced in vivo was 13.9 at a concentration of 3.2 nM (equivalent to 188 ng/mL) with initial cell population around 5.0×10^7 CFU/mL for 1 h [36]. Taken together, our data suggests that the cell-killing activity of cell-free synthesized colicin E3 can be drastically improved by addition or co-expression of the colicin E3 immunity protein to the CFPS reaction to levels that are comparable to those of colicin E3 produced in cells.

3.5. Colicin E1 Is Rapidly Produced and Remains Stable in CFPS

Because CFPS reactions can be lyophilized, stored without cold-chain, and rehydrated to provide simplified and rapid access to high yields of proteins [46], on-demand or distributed manufacturing is a promising application area for CFPS [47]. Production of native or engineered colicins in vitro can be applied to quickly kill a wide variety of pathogens common in resource-limited settings. However, speed of production and protein stability will be critical parameters for these applications and there is currently little information regarding the kinetics of colicin synthesis in CFPS.

In our previous study, colicin Ia produced in CFPS reached 80% of its maximum concentration after just 1 h, reached its maximum concentration at 3 h, and then maintained approximately the same concentration over the remainder of the 20 h incubation time [9]. To investigate earlier time-points and see if such rapid colicin production can be applied to other colicins, we produced colicin E1 in CFPS with 200 ng of linear PCR template and monitored soluble colicin E1 yield over time using ^{14}C-Leu scintillation counting (Figure 3A). Kanamycin was added to the cell-free reaction to a final concentration of 833 µg/mL and the reactions were flash frozen on liquid nitrogen to immediately stop translation after the designated time. Consistent with colicin Ia production [9], soluble colicin E1 was produced rapidly at early time points (until 3 h when it reached its highest concentration) and then maintained approximately the same concentration over the time course (until 20 h). We also performed cell viability assays by treating an initial K361 indicator cell population of 5.0×10^8 CFU/mL with 250 ng/mL of colicin E1 produced in CFPS reactions incubated for various amounts of time. The cell-free produced colicin E1 exhibited very high killing activity and the activities were similar across various CFPS incubation times (Figure 3B), indicating that active colicin E1 is produced rapidly in the CFPS reaction and that up to 20 h, increased incubation time in the CFPS reaction has a negligible effect on colicin E1 cytotoxicity. Taken together with the previous study [9], these results indicate that active colicins can be produced in just 30 mins, reach their maximum yields at approximately 3 h, and that longer incubation times in the CFPS reactions do not adversely affect the activity of colicins.

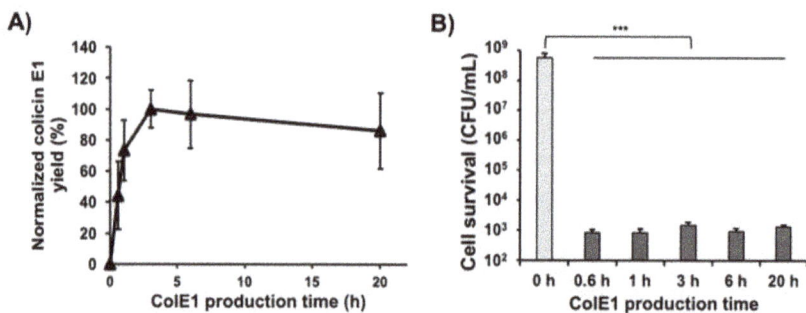

Figure 3. Cell-free production kinetics of colicin E1. (**A**) Yields of soluble colicin E1 produced in CFPS over time. Yields were determined by incorporation of radioactive ^{14}C-leucine and normalized to the maximum yield observed after 3 h incubation. (**B**) Viability of K361 cells (initial cell density 5×10^7 CFU/mL) upon treatment with 250 ng/mL of soluble cell-free produced colicin E1 at 37 °C 220 rpm for 1 h. Error bars indicate standard deviation from two independent cultures with three plating replicates each. *** represents significant difference compared to 0 h sample under the same media conditions with *p*-value < 0.0001.

The rapid and robust production of colicin E1 reported here supports further research into the possible utility of CFPS production of colicins for on-demand protein manufacturing. Because colicins kill non-host *E. coli* cells by recognizing receptors on the cell surface [13], they may be effective in combating *E. coli* strains which cause infectious diseases including urinary tract infection, sepsis/meningitis, and enteric/diarrheal disease [48]. Although effective dosages of colicin E1 in humans have not been determined, a rough estimation of required dosages also suggests that the small-scale production of single dosages of colicin E1 by CFPS is feasible. Fluoroquinolones or third-generation cephalosporins are commonly used to treat many bacterial infections, including those listed above [49]. The FDA has reported that the antibiotic ciprofloxacin (a type of fluoroquinolone) with a MIC_{90} (minimum inhibitory concentration that kills 90% of cell population) of 1.0 µg/mL requires an oral dosing of 250~750 mg per 12 h [50]. Using ciprofloxacin as a guide, the amount of colicin E1, which has a calculated MIC_{90} of 0.016 µg/mL [9], required for a single oral dose with equivalent efficacy can be estimated to be approximately 0.4–1.2 mg. We have found that colicin E1 can be produced at 0.3 mg/mL using CFPS. Therefore, the amount of CFPS reaction volume required to make 1 mg of colicin E1 (approximately one oral dose) would be 3.3 mL, a reasonable volume for synthesis and dosing. Previous studies have shown that total protein yields in *E. coli* CFPS reactions can reach 2.3 mg/mL for sfGFP [51], indicating that additional optimization of CFPS conditions could further decrease the volume of CFPS required for on-demand protein manufacturing of colicins.

4. Conclusions

In summary, we utilized the open nature and engineering flexibility of CFPS to improve colicin M solubility with chaperone-enriched cell extracts, enhance the activity of colicin E3 via co-expression of the immunity protein, and assess the production kinetics and stability of colicin E1 in CFPS reactions. We anticipate that the approaches presented in this study can be applied to optimize the production of other colicins or colicin-like bacteriocins in vitro, providing a useful alternative to the in vivo production of toxic proteins.

Supplementary Materials: The following are available online at http://www.mdpi.com/2409-9279/2/2/28/s1, Table S1: Primers used in this study; Table S2: DNA sequences of all colicin genes; Figure S1: Overall CFPS capacity of chaperone-enriched cell extracts; Figure S2: Effect of additives in colicin E3 yield.

Author Contributions: Conceptualization, methodology, data curation, writing the manuscript: all authors; performed the experiments: X.J. and W.K.; supervision: S.H.H.; funding acquisition: W.K. and S.H.H.

Funding: This work was supported by the National Institute of Allergy and Infectious Diseases of the National Institutes of Health (R15AI130988). This material is based upon work supported by the National Science Foundation (Graduate Research Fellowship under Grant No. DGE-1324585).

Acknowledgments: We thank Willian A. Cramer (Purdue University) for the comments on the manuscript.

Conflicts of Interest: The authors declare no conflicts of interest.

References

1. Blair, J.M.A.; Webber, M.A.; Baylay, A.J.; Ogbolu, D.O.; Piddock, L.J.V. Molecular mechanisms of antibiotic resistance. *Nat. Rev. Microbiol.* **2015**, *13*, 42–51. [CrossRef]
2. Deak, D.; Outterson, K.; Powers, J.H.; Kesselheim, A.S. Progress in the fight against multidrug-resistant bacteria? a review of U.S. Food and Drug Administration–approved antibiotics, 2010–2015. *Ann. Intern. Med.* **2016**, *165*, 363–372. [CrossRef]
3. Cotter, P.D.; Ross, R.P.; Hill, C. Bacteriocins—A viable alternative to antibiotics? *Nat. Rev. Microbiol.* **2013**, *11*, 95–105. [CrossRef]
4. Bahar, A.A.; Ren, D. Antimicrobial peptides. *Pharmaceuticals* **2013**, *6*, 1543–1575. [CrossRef]
5. Lee, J.Y.; Bang, D. Challenges in the chemical synthesis of average sized proteins: Sequential vs. convergent ligation of multiple peptide fragments. *Biopolymers* **2010**, *94*, 441–447. [CrossRef]
6. Salehi, A.S.M.; Smith, M.T.; Bennett, A.M.; Williams, J.B.; Pitt, W.G.; Bundy, B.C. Cell-free protein synthesis of a cytotoxic cancer therapeutic: Onconase production and a just-add-water cell-free system. *Biotechnol. J.* **2016**, *11*, 274–281. [CrossRef] [PubMed]
7. Orth, J.H.C.; Schorch, B.; Boundy, S.; Ffrench-Constant, R.; Kubick, S.; Aktories, K. Cell-free synthesis and characterization of a novel cytotoxic pierisin-like protein from the cabbage butterfly *Pieris rapae*. *Toxicon* **2011**, *57*, 199–207. [CrossRef] [PubMed]

8. Martemyanov, K.A.; Shirokov, V.A.; Kurnasov, O.V.; Gudkov, A.T.; Spirin, A.S. Cell-free production of biologically active polypeptides: Application to the synthesis of antibacterial peptide cecropin. *Protein Expr. Purif.* **2001**, *21*, 456–461. [CrossRef]

9. Jin, X.; Kightlinger, W.; Kwon, Y.-C.; Hong, S.H. Rapid production and characterization of antimicrobial colicins using *Escherichia coli*-based cell-free protein synthesis. *Synth. Biol.* **2018**, *3*, ysy004. [CrossRef]

10. Jin, X.; Hong, S.H. Cell-free protein synthesis for producing 'difficult-to-express' proteins. *Biochem. Eng. J.* **2018**, *138*, 156–164. [CrossRef]

11. Liu, W.-Q.; Zhang, L.; Chen, M.; Li, J. Cell-free protein synthesis: Recent advances in bacterial extract sources and expanded applications. *Biochem. Eng. J.* **2019**, *141*, 182–189. [CrossRef]

12. Bundy, B.C.; Hunt, J.P.; Jewett, M.C.; Swartz, J.R.; Wood, D.W.; Frey, D.D.; Rao, G. Cell-free biomanufacturing. *Curr. Opin. Chem. Eng.* **2018**, *22*, 177–183. [CrossRef]

13. Cascales, E.; Buchanan, S.K.; Duché, D.; Kleanthous, C.; Lloubès, R.; Postle, K.; Riley, M.; Slatin, S.; Cavard, D. Colicin biology. *Microbiol. Mol. Biol. Rev.* **2007**, *71*, 158–229. [CrossRef]

14. El Ghachi, M.; Bouhss, A.; Barreteau, H.; Touzé, T.; Auger, G.; Blanot, D.; Mengin-Lecreulx, D. Colicin M exerts its bacteriolytic effect via enzymatic degradation of undecaprenyl phosphate-linked peptidoglycan precursors. *J. Biol. Chem.* **2006**, *281*, 22761–22772. [CrossRef]

15. Zakharov, S.D.; Wang, X.S.; Cramer, W.A. The colicin E1 TolC-binding conformer: Pillar or pore function of TolC in colicin import? *Biochemistry* **2016**, *55*, 5084–5094. [CrossRef]

16. Jakes, K.S.; Finkelstein, A. The colicin Ia receptor, Cir, is also the translocator for colicin Ia. *Mol. Microbiol.* **2010**, *75*, 567–578. [CrossRef]

17. Sharma, O.; Yamashita, E.; Zhalnina, M.V.; Zakharov, S.D.; Datsenko, K.A.; Wanner, B.L.; Cramer, W.A. Structure of the complex of the colicin E2 R-domain and its BtuB receptor. The outer membrane colicin translocon. *J. Biol. Chem.* **2007**, *282*, 23163–23170. [CrossRef]

18. Cramer, W.A.; Sharma, O.; Zakharov, S.D. On mechanisms of colicin import: The outer membrane quandary. *Biochem. J.* **2018**, *475*, 3903–3915. [CrossRef]

19. González-Montalbán, N.; García-Fruitós, E.; Villaverde, A. Recombinant protein solubility—Does more mean better? *Nat. Biotechnol.* **2007**, *25*, 718–720. [CrossRef]

20. Kim, Y.E.; Hipp, M.S.; Bracher, A.; Hayer-Hartl, M.; Ulrich Hartl, F. Molecular chaperone functions in protein folding and proteostasis. *Annu. Rev. Biochem.* **2013**, *82*, 323–355. [CrossRef]

21. Rosano, G.L.; Ceccarelli, E.A. Recombinant protein expression in *Escherichia coli*: Advances and challenges. *Front. Microbiol.* **2014**, *5*, 172. [CrossRef]

22. Hartl, F.U.; Bracher, A.; Hayer-Hartl, M. Molecular chaperones in protein folding and proteostasis. *Nature* **2011**, *475*, 324–332. [CrossRef]

23. Niwa, T.; Kanamori, T.; Ueda, T.; Taguchi, H. Global analysis of chaperone effects using a reconstituted cell-free translation system. *Proc. Natl. Acad. Sci. USA* **2012**, *109*, 8937–8942. [CrossRef]

24. Stech, M.; Kubick, S. Cell-free synthesis meets antibody production: A review. *Antibodies* **2015**, *4*, 12–33. [CrossRef]

25. Fink, A.L. Chaperone-mediated protein folding. *Physiol. Rev.* **1999**, *79*, 425–449. [CrossRef]

26. Oh, I.-S.; Lee, J.-C.; Lee, M.; Chung, J.; Kim, D.-M. Cell-free production of functional antibody fragments. *Bioprocess Biosyst. Eng.* **2010**, *33*, 127–132. [CrossRef]

27. Kang, S.-H.; Kim, D.-M.; Kim, H.-J.; Jun, S.-Y.; Lee, K.-Y.; Kim, H.-J. Cell-free production of aggregation-prone proteins in soluble and active forms. *Biotechnol. Prog.* **2005**, *21*, 1412–1419. [CrossRef]

28. Gopal, G.J.; Kumar, A. Strategies for the production of recombinant protein in *Escherichia coli*. *Protein J.* **2013**, *32*, 419–425. [CrossRef]

29. Ahn, J.-H.; Chu, H.-S.; Kim, T.-W.; Oh, I.-S.; Choi, C.-Y.; Hahn, G.-H.; Park, C.-G.; Kim, D.-M. Cell-free synthesis of recombinant proteins from PCR-amplified genes at a comparable productivity to that of plasmid-based reactions. *Biochem. Biophys. Res. Commun.* **2005**, *338*, 1346–1352. [CrossRef]

30. Jakes, K.S. Translocation trumps receptor binding in colicin entry into *Escherichia coli*. *Biochem. Soc. Trans.* **2012**, *40*, 1443–1448. [CrossRef]

31. Kwon, Y.-C.; Jewett, M.C. High-throughput preparation methods of crude extract for robust cell-free protein synthesis. *Sci. Rep.* **2015**, *5*, 8663. [CrossRef]

32. Jakes, K.S. The colicin E1 TolC box: Identification of a domain required for colicin E1 cytotoxicity and TolC binding. *J. Bacteriol.* **2017**, *199*, e00412-16. [CrossRef]

Methods Protoc. **2019**, *2*, 28

33. Hong, S.H.; Ntai, I.; Haimovich, A.D.; Kelleher, N.L.; Isaacs, F.J.; Jewett, M.C. Cell-free protein synthesis from a release factor 1 deficient *Escherichia coli* activates efficient and multiple site-specific non-standard amino acid incorporation. *ACS Synth. Biol.* **2014**, *3*, 398–409. [CrossRef]

34. Soelaiman, S.; Jakes, K.; Wu, N.; Li, C.; Shoham, M. Crystal structure of colicin E3: Implications for cell entry and ribosome inactivation. *Mol. Cell* **2001**, *8*, 1053–1062. [CrossRef]

35. Swartz, J.R.; Jewett, M.C.; Woodrow, K.A. Cell-free protein synthesis with prokaryotic combined transcription-translation. *Methods Mol. Biol.* **2004**, *267*, 169–182.

36. Sharma, O.; Cramer, W.A. Minimum length requirement of the flexible N-terminal translocation subdomain of colicin E3. *J. Bacteriol.* **2007**, *189*, 363–368. [CrossRef]

37. Pédelacq, J.-D.; Cabantous, S.; Tran, T.; Terwilliger, T.C.; Waldo, G.S. Engineering and characterization of a superfolder green fluorescent protein. *Nat. Biotechnol.* **2006**, *24*, 79–88. [CrossRef]

38. Acebrón, S.P.; Fernández-Sáiz, V.; Taneva, S.G.; Moro, F.; Muga, A. DnaJ recruits DnaK to protein aggregates. *J. Biol. Chem.* **2008**, *283*, 1381–1390. [CrossRef]

39. Hayer-Hartl, M.; Bracher, A.; Hartl, F.U. The GroEL-GroES chaperonin machine: A nano-cage for protein folding. *Trends Biochem. Sci.* **2016**, *41*, 62–76. [CrossRef]

40. Choi, G.-H.; Lee, D.-H.; Min, W.-K.; Cho, Y.-J.; Kweon, D.-H.; Son, D.-H.; Park, K.; Seo, J.-H. Cloning, expression, and characterization of single-chain variable fragment antibody against mycotoxin deoxynivalenol in recombinant *Escherichia coli*. *Protein Expr. Purif.* **2004**, *35*, 84–92. [CrossRef]

41. Vera, A.; González-Montalbán, N.; Arís, A.; Villaverde, A. The conformational quality of insoluble recombinant proteins is enhanced at low growth temperatures. *Biotechnol. Bioeng.* **2007**, *96*, 1101–1106. [CrossRef]

42. Jewett, M.C.; Fritz, B.R.; Timmerman, L.E.; Church, G.M. *In vitro* integration of ribosomal RNA synthesis, ribosome assembly, and translation. *Mol. Syst. Biol.* **2013**, *9*, 678. [CrossRef]

43. Ng, C.L.; Lang, K.; Meenan, N.A.G.; Sharma, A.; Kelley, A.C.; Kleanthous, C.; Ramakrishnan, V. Structural basis for 16S ribosomal RNA cleavage by the cytotoxic domain of colicin E3. *Nat. Struct. Mol. Biol.* **2010**, *17*, 1241–1246. [CrossRef]

44. Salehi, A.S.M.; Yang, S.-O.; Earl, C.C.; Shakalli Tang, M.J.; Porter Hunt, J.; Smith, M.T.; Wood, D.W.; Bundy, B.C. Biosensing estrogenic endocrine disruptors in human blood and urine: A RAPID cell-free protein synthesis approach. *Toxicol. Appl. Pharmacol.* **2018**, *345*, 19–25. [CrossRef]

45. Zakharov, S.D.; Zhalnina, M.V.; Sharma, O.; Cramer, W.A. The colicin E3 outer membrane translocon: Immunity protein release allows interaction of the cytotoxic domain with OmpF porin. *Biochemistry* **2006**, *45*, 10199–10207. [CrossRef]

46. Smith, M.T.; Berkheimer, S.D.; Werner, C.J.; Bundy, B.C. Lyophilized *Escherichia coli*-based cell-free systems for robust, high-density, long-term storage. *Biotechniques* **2014**, *56*, 186–193. [CrossRef]

47. Hunt, J.P.; Yang, S.O.; Wilding, K.M.; Bundy, B.C. The growing impact of lyophilized cell-free protein expression systems. *Bioengineered* **2017**, *8*, 325–330. [CrossRef]

48. Nataro, J.P.; Kaper, J.B. Diarrheagenic *Escherichia coli*. *Clin. Microbiol. Rev.* **1998**, *11*, 142–201. [CrossRef]

49. Bidell, M.R.; Palchak, M.; Mohr, J.; Lodise, T.P. Fluoroquinolone and third-generation-cephalosporin resistance among hospitalized patients with urinary tract infections due to *Escherichia coli*: Do rates vary by hospital characteristics and geographic region? *Antimicrob. Agents Chemother.* **2016**, *60*, 3170–3173. [CrossRef]

50. *CIPRO®* (*Ciprofloxacin Hydrochloride*) *[Product Information]*; Bayer HealthCare Pharmaceutical Inc.: Whippany, NJ, USA, 1987.

51. Caschera, F.; Noireaux, V. Synthesis of 2.3 mg/ml of protein with an all *Escherichia coli* cell-free transcription–translation system. *Biochimie* **2014**, *99*, 162–168. [CrossRef]

methods and protocols

MDPI

Benchmark

Towards On-Demand *E. coli*-Based Cell-Free Protein Synthesis of Tissue Plasminogen Activator

Seung-Ook Yang [1], Gregory H. Nielsen [1], Kristen M. Wilding [1], Merideth A. Cooper [2], David W. Wood [2] and Bradley C. Bundy [1],*

[1] Department of Chemical Engineering, Brigham Young University, Provo, UT 84602, USA; seungook81@gmail.com (S.-O.Y.); gregnielsen0@gmail.com (G.H.N.); kmwilding13@gmail.com (K.M.W.)

[2] Department of Chemical and Biomolecular Engineering, Ohio State University, Columbus, OH 43210, USA; merideth.cooper@gmail.com (M.A.C.); wood.750@osu.edu (D.W.W.)

* Correspondence: bundy@byu.edu

Received: 2 March 2019; Accepted: 18 April 2019; Published: 21 June 2019

Abstract: Stroke is the leading cause of death with over 5 million deaths worldwide each year. About 80% of strokes are ischemic strokes caused by blood clots. Tissue plasminogen activator (tPa) is the only FDA-approved drug to treat ischemic stroke with a wholesale price over $6000. tPa is now off patent although no biosimilar has been developed. The production of tPa is complicated by the 17 disulfide bonds that exist in correctly folded tPA. Here, we present an *Escherichia coli*-based cell-free protein synthesis platform for tPa expression and report conditions which resulted in the production of active tPa. While the activity is below that of commercially available tPa, this work demonstrates the potential of cell-free expression systems toward the production of future biosimilars. The *E. coli*-based cell-free system is increasingly becoming an attractive platform for low-cost biosimilar production due to recent developments which enable production from shelf-stable lyophilized reagents, the removal of endotoxins from the reagents to prevent the risk of endotoxic shock, and rapid on-demand production in hours.

Keywords: cell-free protein synthesis; CFPS; tPa; tissue plasminogen activator; ischemic stroke

1. Introduction

Pharmaceutical research and development over the past few decades has discovered treatments for a multitude of diseases including stroke [1], cancer [2] and heart disease [3]. Although these innovative treatments have provided much relief, there is still an economic barrier for many patients in need due to the high cost of drug treatments. An example is Tissue plasminogen activator (tPa) which is the only FDA-approved drug to treat ischemic stroke and costs over $6000 [4]. Ischemic stroke is responsible for ~80% of all strokes with strokes annually causing over 5 million deaths and disabling ~50% of the survivors [5]. Fortunately, tPa, stroke education, and access to medical care has greatly reduced the mortality rate in developed countries, while the mortality rate is significantly higher in developing countries where patients cannot afford tPa and have less access to medical education and care [5]. While tPa is now off patent, no generic or biosimilar has been developed. Biosimilars are biological products approved for usage based on their highly similar functions with FDA-approved biological products. Biosimilars are beneficial in terms of cost savings and increased patient accessibility. However, biosimilars do not reduce drug prices to the extent that generics of small-molecule drugs do. Biosimilars are only 10–30% less than the original drug price [6]. Even so, it is estimated that biosimilar introduction into the market will save the U.S. up to $378 billion over the next 20 years [7]. Thus, there is an unmet technological need to produce biosimilar therapeutics at extremely low costs to enable developing countries have access to life-saving therapeutics.

Escherichia coli-based cell-free protein synthesis (CFPS) is increasingly becoming an attractive platform as a future low-cost biosimilar production. Advantages of *E. coli*-based CFPS include (1) an open reaction environment for greater control over reaction conditions, (2) the ability to lyophilize reagents for shelf-stable storage and stockpiling, (3) on-demand rapid production in response to needs, (4) the ability to express proteins in an endotoxin-free environment, (5) scalability from the microliter to 100 liter scale, (6) very low cost reagents, (7) PEGylation optimization [8], and (7) the ability to automate the protein production process [9–18]. The production of active forms of multiple therapeutic proteins has been demonstrated in recent years, targeting a variety of diseases [19,20]. Indeed, a recent report demonstrated the *E. coli*-based CFPS production of a biosimilar to the FDA-approved crisantaspase [21]. Here, we further assess CFPS as a tool towards the production of biosimilars by demonstrating the production of a highly disulfide-bonded therapeutic protein, tPa.

tPA is a serine protease and functions by activating plasminogen to dissolve blood plasma proteins found in blood clots. tPa has a total of 35 cysteine residues which allow the formation of 17 disulfide bonds [22–24], and consists of 527 amino acids with five domains (such as Kringle I, Finger, EGF, thrombolytic Kringle II, and protease domains) [25]. Currently, tPA is produced in Chinese hamster ovary (CHO) cells and is post-translationally glycosylated [26]. Other eukaryotic cell expression techniques suffer due to the product not folding correctly, transport limitations, and hyperglycosylation [22,27–29]. Part of the difficulty in producing tPA is the 17 disulfide bonds required for the correctly folded protein. Attempts to produce tPA in periplasm of *E. coli* have suffered from rare codon usage, mis-folding, and loss of function [23,24]. An engineered *E. coli* (Origami B, DE3 strain), where two thioredoxins (TrxA and TrxC) and three glutaredoxins were oxidized, was used to express a full length of tPA. However, the expression yields and the specific activity were not reported [22]. Here, we demonstrate for the first time the expression of active tPA in an *E. coli*-based CFPS environment using a glutathione buffer and prokaryotic disulfide bond isomerase C (DsbC) [30]. tPA expression methods with CFPS, the specific activity of tPa, and demonstration of produced tPa's ability to lyse a human blood clot is described.

2. Materials and Methods

2.1. Extract Preparation

The *E. coli* extract was prepared using BL21 Star^{TM} (DE3) *E. coli* strain purchased from Invitrogen (Carlsbad, CA, USA). This strain harbored pOFX-GroEL/ES and pET40-DsbC, key components to the process. A starting culture of this *E. coli* BL21 Star^{TM} (DE3) strain grew in 5 mL of LB media (supplemented with 100 µg/mL spectinomycin and kanamycin) overnight at 37 °C with shaking at 280 rpm. The culture was transferred to 100 mL of LB media (supplemented with 100 µg/mL spectinomycin and kanamycin) and grown until reaching an OD600 reading of 2.0. The 105 mL culture was then transferred to 1 L of media in a Tunair flask. Then, 1 mM of Isopropyl β-D-1-thiogalactopyranoside (IPTG) was added upon reaching 0.6 of OD600 to overexpress T7 RNA polymerase. Cells were harvested at the end of the exponential growth phase (OD600 of 1.2) using centrifugation at 6000 RCF for 10 min at 4 °C. Cells were washed with chilled Buffer A (10 mM Tris-acetate pH 8.2, 14 mM magnesium acetate, 60 mM potassium glutamate, and 1 mM dithiothreitol (DTT)) by centrifugation at 6000 RCF for 10 min at 4 °C. Cells were re-suspended with the buffer A (g of cell/mL of buffer A), and protein synthesis machinery was extracted using an EmulsiFlex French Press homogenizer (Avestin, Ottowa, ON, Canada) at 20,000 psi. The homogenized cells were centrifuged at 12,000 RCF for 30 min at 4 °C to clear the lysate. The supernatant was incubated in a shaking incubator for 30 min at 280 rpm and 37 °C. The extract was flash-frozen in liquid nitrogen for one min before being stored at −80 °C.

2.2. Cell-Free Protein Synthesis

Cell-free protein synthesis of tPA was carried out in 80–100 μL reactions housed in a 2.0-mL Eppendorf tube at 37 °C for 3 h, with PANOxSP as an energy source as reported previously [10]. Expression of tPa was templated using a pET-tPa plasmid which encodes the expression of human tPa under a T7 RNA polymerase promoter. Upon analysis, the codon sequence contained some rare codons based on *E. coli* usage frequencies, but was deemed suitable for initial tPa expression studies. Each reaction contained 25% reaction volume of the cell extract, 1.2 nM plasmid, 12 to 15 mM magnesium glutamate, 1 mM 1,4-Diaminobutane, 1.5 mM Spermidine, 33.3 mM phosphoenolpyruvate (PEP), 10 mM ammonium glutamate, 175 mM potassium glutamate, 2.7 mM potassium oxalate, 0.33 mM nicotinamide adenine dinucleotide (NAD), 0.27 mM coenzyme A (CoA), 1.2 mM ATP, 0.86 mM CTP, 0.86 mM GTP, 0.86 mM UTP, 0.17 mM folinic acid, and 2 mM of all 20 amino acids except glutamic acid. The cell extract had overexpressed GroEL/ES and DsbC as mentioned above. Reactions were supplemented with purified DsbC where specified. Additionally, a 5 mM glutathione buffer (GSSG:GSH = 4:1) was included in the CFPS reaction reagents to support disulfide bond formation.

2.3. Measuring Protein Concentration

Next, 5 μM of radiolabeled (U-14C) Leucine (PerkinElmer, Waltham, MA, USA) was added to the CFPS reaction. A volume of 3 μL of the CFPS reaction was spotted on each of the three separate pieces of Whatman 3MM chromatography paper and dried at 37 °C. Spotted papers were placed in a beaker on ice and covered with 5% (v/v) TCA at 4 °C to precipitate the proteins onto the filter paper for 15 min. The solution was exchanged with fresh TCA three times. Following the TCA washing steps, the papers were dried at 37 °C. The radioactivity of both TCA-precipitated and non-TCA-precipitated samples was measured using a LS6500 Multipurpose Scintillation Counter (Beckman Coulter, Brea, CA, USA). The fraction of incorporated leucine in washed and unwashed protein were used to determine the amount of protein synthesized. Soluble yields were determined by sample centrifugation at 4 °C and 17,000 × *g* for 15 min, followed by TCA-precipitation and scintillation counting of the supernatants.

2.4. tPa Purification

The expressed tPA was dialyzed in NET buffer (150 mM NaCl and 20 mM Tris-HCl, pH 6.7). The cell-free reaction that produced tPA was inserted into a dialysis tubing bag (Spectra/Por, Rancho Dominguez, CA, USA) with a molecular weight cut off of 6–8 kDa. The sample was immediately dialyzed against 300 mL NET buffer for 18 h at 4 °C with three buffer exchanges. Dialyzed samples were loaded directly onto the HisPurTM Ni-NTA spin column (Thermo Scientific, Rockford, IL, USA) that was pre-equilibrated with NET buffer containing 10 mM imidazole. The column was equilibrated with sample proteins for 30 min at 4 °C. The column was washed with a NET buffer containing 10 mM imidazole (1.2 mL). tPA was eluted with a NET buffer containing 500 mM imidazole (300 μL). Eluted samples were concentrated by dialysis in 500 mL of NET buffer containing 40% glycerol.

2.5. tPa Activity Assay

The SensoLyte AMC tPA activity assay kit was purchased from AnaSpec (Fremont, CA, USA). The reaction buffer, AMC substrate, and standard AMC (fluorescence) were provided in the kit. A standard curve was plotted with varying concentration of the standard AMC. Pure tPA was added to 1x reaction buffer that was provided to make a 50-μL enzyme solution. Then, 50 μL of the AMC substrate solution was added directly to the enzyme solution for the activity measurement. Fluorescence was monitored at excitation/emission = 354/442 nm.

2.6. Blood Clot Lysis Assay

Blood for the clot lysis assay was a donation from the BYU Student Health Center. Aliquots of the blood were coagulated in the incubator at 37 °C for three hours. Non-coagulated blood was removed using an aspirator. The weight of the blood was measured before the incubation with tPA enzyme solution (1.9, 0.43, and 0 µM) in 100 µL of water. After 16 h of incubation with the tPA enzyme solution, the non-coagulated part was removed by using the aspirator, and the weight of the blood was measured.

3. Results and Discussion

3.1. tPA Expression and Specific Activity

To facilitate the expression and proper folding of tPA, 5 mM of glutathione buffer (with a 4:1 ratio of oxidized and reduced glutathione) was combined with cell extract containing overexpressed GroEL/ES and DsbC in a CFPS reaction (Figure 1a). Combining the above glutathione buffer with DsbC has previously been shown to enable the high yielding production of disulfide-bonded proteins with CFPS [30]. tPA was expressed at 37 °C for 3 h with CFPS supplemented with DsbC to a final DsbC concentration of 3.2 or 6.5 µM. Higher yields were obtained with 6.5 µM of DsbC, reaching 348 µg/mL of soluble tPa (Figure 1b). As reported previously, adding DsbC has been shown to increase both total and soluble protein yields of disulfide-bonded proteins [19]. However, the extent of total yield increase is higher with tPa. One possible explanation could be a higher propensity of the highly disulfide-bonded tPa to catalyze aggregation when incorrectly folded, which may in turn inhibit transcription/translation machinery. Densitometry results indicate that a total of 30% of the expressed tPa was the full-length protein, meaning the actual total soluble tPa yield is 104 µg/mL (Figure 2a). tPa was purified using Ni affinity chromatography and autoradiogram analysis of C14-Leu radiolabeled tPA showed the presence of two protein bands (Figure 2a). Native tPA possesses a total of three glycosylation sites: Asn 117, 184, and 448 [31]. Studies have shown that native tPA is cleaved by serine proteases to form a disulfide-bond linked two-chain tPA when glycosylation is missing at Asn 184 [32–34]. Serine proteases present in *E. coli* could cleave non-glycosylated tPA [35] with the distinct truncated band suggesting the product was cleaved by location-specific protease. In the future, this truncation could be reduced by mutagenesis to remove serine protease cleavage sites or engineering of glycosylation into the *E. coli*-based CFPS.

Using the SensoLyte AMC tPA activity assay kit [36], the specific activity of purified tPA was determined to be 3000 µmol/min/mg. This specific activity correlates to less than 1% of the reported specific activity of commercially available tPa [24]. The presence of any activity is encouraging considering the 17 correctly linked disulfide bonds among the 35 cystine residues required for activity, the lack of glycosylation, and the significant concentration of truncated tPa product. However, significant optimization is required before a tPa biosimilar from an *E. coli*-based CFPS system could be realized. This includes engineering of CHO-type glycosylation and tuning the redox potential and chaperone concentrations to better facilitate disulfide bond formation. Engineering glycosylation into *E. coli*-based CFPS appears to be the biggest challenge to the production biosimilars that require glycosylation. Fortunately, recent engineering work engineering asparagine-linked glycosylation in *E. coli*-based CFPS suggests engineering CHO-type glycosylation may become a reality sooner than previously anticipated [37].

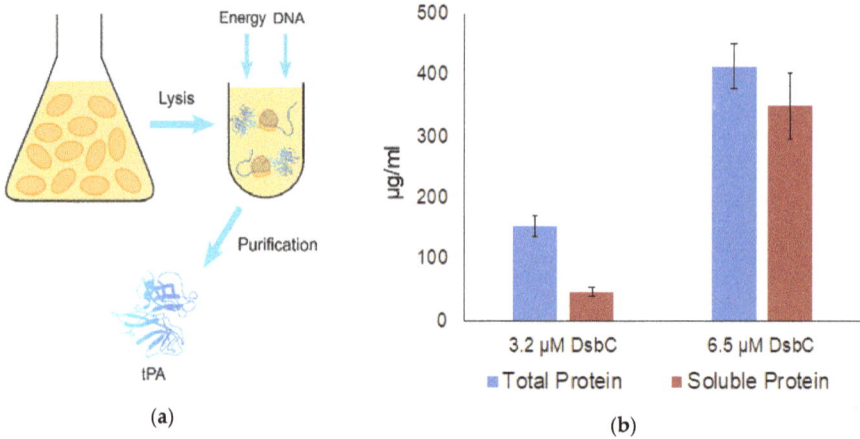

Figure 1. (**a**) Schematic showing the tissue plasminogen activator (tPA) expression process using a cell-free protein synthesis (CFPS) platform. *E. coli* cells are first grown and lysed. Next, cell extract, energy and DNA are all combined to express tPA which is then purified; (**b**) Total and soluble tPA yield with disulfide bond isomerase C (DsbC). Both reactions were incubated at 37 °C for 3 h. The final concentration of DsbC in each reaction is shown. Error bars represent one standard deviation.

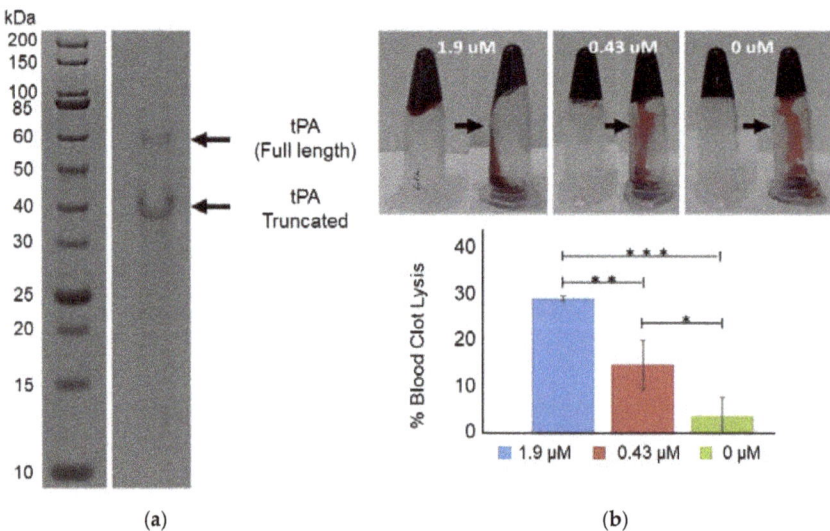

Figure 2. Purification and Activity of tPA. Nickel-affinity column-purified tPA was separated by SDS-PAGE and the gel was exposed to autoradiography film. Samples were also tested with a blood clot lysis assay: (**a**) Autoradiogram analysis showed both a full-length (63 kDa) and truncated tPA (about 40 kDa); (**b**) Analysis of blood clot lysis with CFPS expressed tPA. *** = $p < 0.005$, ** = $p < 0.05$, and * = $p < 0.1$. Error bars represent one standard deviation.

3.2. Blood Clot Lysis

E. coli-based CFPS expressed and purified tPA was also assessed for its ability to lyse human blood clots (Figure 2b). Three samples were prepared with varying tPA concentrations (1.9, 0.43, and 0 µM). Each sample was added to human coagulated blood. After overnight incubation at 37 °C, non-coagulated blood was cleared, and the weight of the remaining blood was measured. About

30% of the coagulated blood was lysed when 1.9 μM tPA was added to the coagulated blood sample. The clot lysis activity of the 1.9 μM tPa samples were statistically greater than the 0.43 μM and 0 μM samples ($p < 0.05$ and $p < 0.005$ respectively). The 0.43 μM tPa samples also appeared to be more active than the control sample with 0 μM tPa with an average of ~15% clot lysis activity compared to an average of ~5%. However, the difference was not as statistically significant ($p < 0.1$). Overall, these results demonstrate that tPa expressed using *E. coli*-based CFPS activated blood clot lysis.

4. Conclusions

Here, we demonstrate for the first time the production of active tPA using *E. coli*-based cell-free protein synthesis. Recent advances using low-cost *E. coli*-based CFPS has increased its fitness to produce both biosimilars and new therapeutic proteins. For example, this system can be used to optimize the site of PEGylation which is an important post-translational modification in the development of 2nd-generation drugs [8]. In addition, this system has very recently been used to produce a biosimilar to FDA-approved crisantaspase [21]. However, the system is currently limited to non-glycosylated therapeutics. Continued research engineering glycosylation pathways into the CFPS system is essential to enabling glycosylated biosimilar production in the future.

Author Contributions: Conceptualization, S.O.Y. and B.C.B.; methodology, S.O.Y. and B.C.B.; validation, K.M.W.; formal analysis, S.O.Y., G.H.N. and K.M.W.; investigation, S.O.Y.; resources, B.C.B., M.A.C., and D.W.D.; writing—original draft preparation, S.O.Y., G.H.N. and B.C.B.; writing—review and editing, G.H.N., M.A.C., D.W.D., K.M.W., and B.C.B.; supervision, B.C.B.; project administration, B.C.B.; funding acquisition, B.C.B.

Funding: This research was funded by the National Science Foundation, grant number 1254148 and a BYU Simmons Center Cancer Research Fellowship.

Acknowledgments: The authors acknowledge the BYU Student Health Center for their support.

Conflicts of Interest: The authors declare no conflict of interest. The funders had no role in the design of the study; in the collection, analyses, or interpretation of data; in the writing of the manuscript, or in the decision to publish the results.

References

1. Gravanis, I.; Tsirka, S.E. Tissue-type plasminogen activator as a therapeutic target in stroke. *Expert Opin. Ther. Targets* **2008**, *12*, 159–170. [CrossRef] [PubMed]
2. Salehi, A.S.M.; Earl, C.C.; Muhlestein, C.; Bundy, B.C. Escherichia coli-based cell-free extract development for protein-based cancer therapeutic production. *Int. J. Dev. Biol.* **2016**, *60*, 237–243. [CrossRef] [PubMed]
3. Senis, Y.A.; Barr, A.J. Targeting Receptor-Type Protein Tyrosine Phosphatases with Biotherapeutics: Is Outside-in Better than Inside-Out? *Molecules* **2018**, *23*. [CrossRef] [PubMed]
4. Boudreau, D.M.; Guzauskas, G.; Villa, K.F.; Fagan, S.C.; Veenstra, D.L. A model of cost-effectiveness of tissue plasminogen activator in patient subgroups 3 to 4.5 hours after onset of acute ischemic stroke. *Ann. Emerg. Med.* **2013**, *61*, 46–55. [CrossRef] [PubMed]
5. Donkor, E.S. Stroke in the 21(st) Century: A Snapshot of the Burden, Epidemiology, and Quality of Life. *Stroke Res. Treat.* **2018**, *2018*, 3238165. [PubMed]
6. Mehr, S.R.; Brook, R.A. Factors influencing the economics of biosimilars in the US. *J. Med. Econ.* **2017**, *20*, 1268–1271. [CrossRef] [PubMed]
7. Cornes, P. The economic pressures for biosimilar drug use in cancer medicine. *Target. Oncol.* **2012**, *7* (Suppl. 1), S57–S67. [CrossRef]
8. Wilding, K.M.; Smith, A.K.; Wilkerson, J.W.; Bush, D.B.; Knotts, T.A.t.; Bundy, B.C. The Locational Impact of Site-Specific PEGylation: Streamlined Screening with Cell-Free Protein Expression and Coarse-Grain Simulation. *ACS Synth. Biol.* **2018**, *7*, 510–521. [CrossRef]
9. Shrestha, P.; Smith, M.T.; Bundy, B.C. Cell-free unnatural amino acid incorporation with alternative energy systems and linear expression templates. *New Biotechnol.* **2014**, *31*, 28–34. [CrossRef]
10. Salehi, A.S.; Smith, M.T.; Bennett, A.M.; Williams, J.B.; Pitt, W.G.; Bundy, B.C. Cell-free protein synthesis of a cytotoxic cancer therapeutic: Onconase production and a just-add-water cell-free system. *Biotechnol. J.* **2016**, *11*, 274–281. [CrossRef]

11. Smith, M.T.; Berkheimer, S.D.; Werner, C.J.; Bundy, B.C. Lyophilized Escherichia coli-based cell-free systems for robust, high-density, long-term storage. *Biotechniques* **2014**, *56*, 186–193. [CrossRef] [PubMed]
12. Smith, M.T.; Wilding, K.M.; Hunt, J.M.; Bennett, A.M.; Bundy, B.C. The emerging age of cell-free synthetic biology. *FEBS Lett.* **2014**, *588*, 2755–2761. [CrossRef] [PubMed]
13. Earl, C.C.; Smith, M.T.; Lease, R.A.; Bundy, B. Polyvinylsulfonic Acid: A Low-cost RNase Inhibitor for Enhanced RNA Preservation and Cell-free Protein Translation. *Bioengineered* **2018**, *9*, 90–97. [CrossRef] [PubMed]
14. Wilding, K.M.; Schinn, S.M.; Long, E.A.; Bundy, B.C. The emerging impact of cell-free chemical biosynthesis. *Curr. Opin. Biotechnol.* **2018**, *53*, 115–121. [CrossRef] [PubMed]
15. Bundy, B.C.; Hunt, J.P.; Jewett, M.C.; Swartz, J.R.; Wood, D.W.; Frey, D.D.; Rao, G. Cell-free biomanufacturing. *Curr. Opin. Chem. Eng.* **2018**, *22*, 177–183. [CrossRef]
16. Hunt, J.P.; Yang, S.O.; Wilding, K.M.; Bundy, B.C. The growing impact of lyophilized cell-free protein expression systems. *Bioengineered* **2017**, *8*, 325–330. [CrossRef] [PubMed]
17. Schinn, S.M.; Bradley, W.; Groesbeck, A.; Wu, J.C.; Broadbent, A.; Bundy, B.C. Rapid in vitro screening for the location-dependent effects of unnatural amino acids on protein expression and activity. *Biotechnol. Bioeng.* **2017**, *114*, 2412–2417. [CrossRef] [PubMed]
18. Salehi, A.S.; Shakalli Tang, M.J.; Smith, M.T.; Hunt, J.M.; Law, R.A.; Wood, D.W.; Bundy, B.C. Cell-Free Protein Synthesis Approach to Biosensing hTRbeta-Specific Endocrine Disruptors. *Anal. Chem.* **2017**, *89*, 3395–3401. [CrossRef]
19. Yang, J.H.; Kanter, G.; Voloshin, A.; Levy, R.; Swartz, J.R. Expression of active murine granulocyte-macrophage colony-stimulating factor in an Escherichia coli cell-free system. *Biotechnol. Progr.* **2004**, *20*, 1689–1696. [CrossRef]
20. Zimmerman, E.S.; Heibeck, T.H.; Gill, A.; Li, X.; Murray, C.J.; Madlansacay, M.R.; Tran, C.; Uter, N.T.; Yin, G.; Rivers, P.J.; et al. Production of Site-Specific Antibody–Drug Conjugates Using Optimized Non-Natural Amino Acids in a Cell-Free Expression System. *Bioconjugate Chem.* **2014**, *25*, 351–361. [CrossRef]
21. Wilding, K.M.; Hunt, J.P.; Wilkerson, J.W.; Funk, P.J.; Swensen, R.L.; Carver, W.C.; Christian, L.M. Endotoxin-free E. coli-based cell-free protein synthesis: Pre-expression endotoxin removal approaches for on-demand cancer therapeutic production. *Biotechnol. J.* **2019**, *14*, 1800271. [CrossRef] [PubMed]
22. Majidzadeh, A.K.; Mahboudi, F.; Hemayatkar, M.; Davami, F.; Barkhordary, F.; Adeli, A.; Soleimani, M.; Davoudi, N.; Khalaj, V. Human Tissue Plasminogen Activator Expression in Escherichia coli using Cytoplasmic and Periplasmic Cumulative Power. *Avicenna J. Med. Biotechnol.* **2010**, *2*, 131–136.
23. Qiu, J.; Swartz, J.R.; Georgiou, G. Expression of active human tissue-type plasminogen activator in Escherichia coli. *Appl. Environ. Microbiol.* **1998**, *64*, 4891–4896. [PubMed]
24. Lee, H.J.; Im, H. Soluble Expression and Purification of Human Tissue-type Plasminogen Activator Protease Domain. *Bull. Korean Chem. Soc.* **2010**, *31*, 2607–2612. [CrossRef]
25. Fathi-Roudsari, M.; Akhavian-Tehrani, A.; Maghsoudi, N. Comparison of Three Escherichia coli Strains in Recombinant Production of Reteplase. *Avicenna J. Med. Biotechnol.* **2016**, *8*, 16–22. [PubMed]
26. Pennica, D.; Holmes, W.E.; Kohr, W.J.; Harkins, R.N.; Vehar, G.A.; Ward, C.A.; Bennett, W.F.; Yelverton, E.; Seeburg, P.H.; Heyneker, H.L.; et al. Cloning and expression of human tissue-type plasminogen activator cDNA in E. coli. *Nature* **1983**, *301*, 214–221. [CrossRef] [PubMed]
27. Furlong, A.M.; Thomsen, D.R.; Marotti, K.R.; Post, L.E.; Sharma, S.K. Active human tissue plasminogen activator secreted from insect cells using a baculovirus vector. *Biotechnol. Appl. Biochem.* **1988**, *10*, 454–464.
28. Martegani, E.; Forlani, N.; Mauri, I.; Porro, D.; Schleuning, W.D.; Alberghina, L. Expression of high levels of human tissue plasminogen activator in yeast under the control of an inducible GAL promoter. *Appl. Microbiol. Biotechnol.* **1992**, *37*, 604–608. [CrossRef]
29. Steiner, H.; Pohl, G.; Gunne, H.; Hellers, M.; Elhammer, A.; Hansson, L. Human tissue-type plasminogen activator synthesized by using a baculovirus vector in insect cells compared with human plasminogen activator produced in mouse cells. *Gene* **1988**, *73*, 449–457. [CrossRef]
30. Goerke, A.R.; Swartz, J.R. Development of cell-free protein synthesis platforms for disulfide bonded proteins. *Biotechnol. Bioeng.* **2008**, *99*, 351–367. [CrossRef]
31. Pohl, G.; Kallstrom, M.; Bergsdorf, N.; Wallen, P.; Jornvall, H. Tissue plasminogen activator: peptide analyses confirm an indirectly derived amino acid sequence, identify the active site serine residue, establish glycosylation sites, and localize variant differences. *Biochemistry* **1984**, *23*, 3701–3707. [CrossRef] [PubMed]

32. Rajapakse, S.; Ogiwara, K.; Takano, N.; Moriyama, A.; Takahashi, T. Biochemical characterization of human kallikrein 8 and its possible involvement in the degradation of extracellular matrix proteins. *FEBS Lett.* **2005**, *579*, 6879–6884. [CrossRef] [PubMed]

33. Rijken, D.C.; Hoylaerts, M.; Collen, D. Fibrinolytic properties of one-chain and two-chain human extrinsic (tissue-type) plasminogen activator. *J. Biol. Chem.* **1982**, *257*, 2920–2925.

34. Wallen, P.; Pohl, G.; Bergsdorf, N.; Ranby, M.; Ny, T.; Jornvall, H. Purification and characterization of a melanoma cell plasminogen activator. *Eur. J. Biochem.* **1983**, *132*, 681–686. [CrossRef]

35. Strongin, A.Y.; Gorodetsky, D.I.; Stepanov, V.M. The study of Escherichia coli proteases. Intracellular serine protease of E. coli-an analogue of bacillus proteases. *J. Gen. Microbiol.* **1979**, *110*, 443–451. [CrossRef] [PubMed]

36. Rodier, M.; Prigent-Tessier, A.; Bejot, Y.; Jacquin, A.; Mossiat, C.; Marie, C.; Garnier, P. Exogenous t-PA Administration Increases Hippocampal Mature BDNF Levels. Plasmin- or NMDA-Dependent: Mechanism? *PLoS ONE* **2014**, *9*, e92416. [CrossRef] [PubMed]

37. Jaroentomeechai, T.; Stark, J.C.; Natarajan, A.; Glasscock, C.J.; Yates, L.E.; Hsu, K.J.; Mrksich, M.; Jewett, M.C.; DeLisa, M.P. Single-pot glycoprotein biosynthesis using a cell-free transcription-translation system enriched with glycosylation machinery. *Nat. Commun.* **2018**, *9*, 2686. [CrossRef] [PubMed]

methods and protocols

MDPI

Article

Accelerating the Production of Druggable Targets: Eukaryotic Cell-Free Systems Come into Focus

Lena Thoring, Anne Zemella, Doreen Wüstenhagen and Stefan Kubick *

Fraunhofer Institute for Cell Therapy and Immunology (IZI), Branch Bioanalytics and Bioprocesses (IZI-BB), Am Mühlenberg 13, D-14476 Potsdam, Germany; lena.thoring@izi-bb.fraunhofer.de (L.T.); anne.zemella@izi-bb.fraunhofer.de (A.Z.); doreen.wuestenhagen@izi-bb.fraunhofer.de (D.W.)
* Correspondence: Stefan.Kubick@izi-bb.Fraunhofer.de

Received: 15 February 2019; Accepted: 10 April 2019; Published: 16 April 2019

Abstract: In the biopharmaceutical pipeline, protein expression systems are of high importance not only for the production of biotherapeutics but also for the discovery of novel drugs. The vast majority of drug targets are proteins, which need to be characterized and validated prior to the screening of potential hit components and molecules. A broad range of protein expression systems is currently available, mostly based on cellular organisms of prokaryotic and eukaryotic origin. Prokaryotic cell-free systems are often the system of choice for drug target protein production due to the simple generation of expression hosts and low cost of preparation. Limitations in the production of complex mammalian proteins appear due to inefficient protein folding and posttranslational modifications. Alternative protein production systems, so-called eukaryotic cell-free protein synthesis systems based on eukaryotic cell-lysates, close the gap between a fast protein generation system and a high quality of complex mammalian proteins. In this study, we show the production of druggable target proteins in eukaryotic cell-free systems. Functional characterization studies demonstrate the bioactivity of the proteins and underline the potential for eukaryotic cell-free systems to significantly improve drug development pipelines.

Keywords: in vitro translation; protein production; drug development; cell-free protein synthesis; eukaryotic lysates; microsomes; growth factors; enzymes

---

1. Introduction

Over the last decades, the pharmaceutical market has changed its focus from the production of small molecule components to more complex biopharmaceuticals. Biotherapeutics gain a growing share of the global pharmaceutical market [1] and newly developed products are based on therapeutical proteins such as monoclonal antibodies [2]. The rising importance of protein molecules for the pharmaceutical industry is not only present in the topic of biopharmaceuticals, but also in the case of drug development. Drug discovery and development is a costly and time-consuming process, which requires typically 12–15 years from an original idea to the launch of a final product [3]. Therefore one of the challenges in the pharmaceutical industry is to obtain optimal and efficient drug discovery pipelines [4]. The first step of drug discovery comprises the identification of a drug target and the vast majority of approved drugs are proteins [5]. Major drug target classes belong to antineoplastics, G protein-coupled receptors (GPCR's), ion channels, kinases and proteases [5]. Following the process of identification, a target is further validated and possible hits are identified in a so-called lead discovery phase. The target validation techniques range from in vitro tools to knock down and knock out of the gene of interest to animal models mimicking the desired disease [6]. Current approaches for the validation are often time-consuming and cost inefficient [7]. Apart from this, the identification of hit molecules interacting with target proteins is based on screening approaches where a variety of efficient strategies exist [3]. A recombinant protein is often a prerequisite for most drug development strategies and this target protein needs to be produced in reliable systems. Nowadays, a huge number of protein production systems are available and used for drug discovery to evaluate the structure of disease-associated proteins and respective protein-protein interactions. Most of these systems are based on genetically engineered cell lines originating from pro- and eukaryotes. One of the best platforms for the production of soluble proteins is, for example, the pCold-glutathione S-transferase (GST) system, which is based on the prokaryotic expression host *Escherichia coli* [8]. Limitations of prokaryotic systems occur when complex mammalian target proteins requiring posttranslational modifications, cofactors and chaperons for correct protein folding, assembly and activity need to be produced. To circumvent these issues, eukaryotic cell-based expression systems are available, including yeast systems (*Pichia pastoris, Saccharomyces cerevisiae, Kluyveromyces lactis*) and mammalian systems (HEK293, Chinese hamster ovary cells (CHO cells)). Interestingly, mammalian systems are only rarely reported in screening literature [9]. Generation of eukaryotic stable cell lines for protein production purposes can be quite laborious as cells grow slowly, production time increases and protein yields are low thereby leading to costly protein production processes. Alternative protein production platforms, so-called cell-free protein synthesis systems were developed based on cell lysates instead of a whole, living cell. This technology emerged as a powerful and flexible tool for fast and efficient production of proteins whereby the surrounding environment can be easily adapted for the required approach [10]. Cell-free systems originating from different cellular hosts are currently available. The most commonly used cell-free system is based on *E. coli* cell lysates [11,12], which typically achieve up to 1 mg/mL of de novo synthesized protein. This system is already used for screening approaches in terms of the development of protein in situ arrays (PISA) [13] as well as nucleic acid programmable protein arrays (NAPPA) [14,15]. *E. coli* cell-free systems are limited in the performance of posttranslational modifications. Therefore, such systems are not suitable for the synthesis of complex mammalian proteins. This led to the development of the first eukaryotic cell-free protein synthesis system originating from rabbit reticulocytes. The rabbit reticulocyte system showed low translation efficiencies and posttranslational modifications of proteins can only be conducted by supplementing exogenous microsomes [16,17]. Over the years, a broad range of eukaryotic cell-free systems was developed exhibiting improved translational efficiencies and the opportunity to produce complex mammalian proteins due to the presence of endogenous microsomes [18]. Apart from the wheat germ cell-free system, which is characterized by a highly efficient translational machinery but limited in posttranslational modifications, eukaryotic cell-free systems based on yeast [19,20], insect cells [21], CHO cells [22], tobacco cells [23] and human cell lines [24] harbor endogenous microsomes. These microsomes are derived from the endoplasmic reticulum, thereby enabling a co-translational translocation of proteins and ER-based posttranslational modifications such as glycosylation, disulfide bridging and lipidation.

Despite significant advantages to produce challenging mammalian proteins, which the eukaryotic cell-free systems provide, they are typically not part of the drug discovery pipeline until now. In the past, eukaryotic cell-free systems were mostly cost-ineffective and characterized by low productivities, which made the technology inefficient for industrial applications. Immense development in the area of extract preparation, system optimization and reduction of process costs lead to well-established eukaryotic cell-free systems nowadays suitable for industrial applications. In this study, we demonstrate the production of druggable targets in eukaryotic cell-free systems. The general principle of these systems and future applications are pointed out in Figure 1. We started to produce the target proteins based on linear DNA templates and plasmids, transcribed DNA into mRNA in an in vitro transcription step using T7 RNA polymerase and used different eukaryotic cell-free systems for the production of the required proteins. The produced drug target proteins were functionally characterized to show a proof of concept for the application of the platform technology to the drug development pipeline.

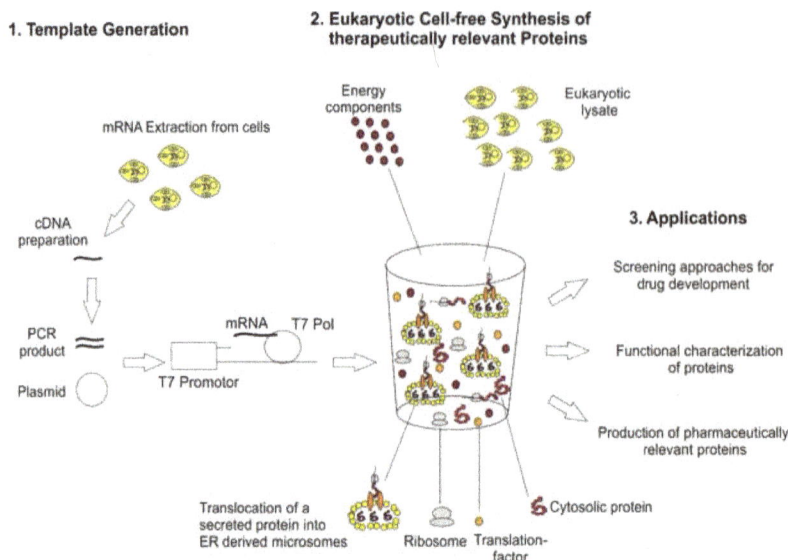

Figure 1. General principle of eukaryotic cell-free technology for research and therapeutical applications. For eukaryotic cell-free protein synthesis, a suitable DNA template is required, which can be directly prepared from cellular mRNA by RT-PCR. In this way, 5′ and 3′ regulatory sequences (T7 promotor, T7 terminator, stem loops and hairpin sequences) are added to the DNA template. Alternatively, plasmids harboring regulatory sequences can be used for eukaryotic cell-free protein synthesis. The DNA template is transcribed into mRNA using T7 RNA polymerase (T7 Pol) directly added to the cell-free synthesis reaction. Eukaryotic cell-free protein synthesis is based on a eukaryotic cell lysate including endogenous microsomes derived from endoplasmic reticulum. Special eukaryotic lysates like wheat germ and rabbit reticulocyte lysate do not include endogenous microsomes. The eukaryotic cell lysate is supplemented with previously produced mRNA and buffer and energy components to perform cell-free protein synthesis. Applications of eukaryotic cell-free protein synthesis are the development of novel screening platforms for drugs, the functional characterization of proteins and the production of biotherapeutics.

2. Materials and Methods

2.1. Methods for Protein Production

This section comprises all methods and materials required for eukaryotic cell-free protein synthesis and a detailed analysis of the translational activity in eukaryotic cell-free systems. Various eukaryotic cell-free systems were applied and most of the results are derived in transcription/translation linked cell-free systems where transcription of mRNA and translation of a target protein are separated processes. The generation and design of required DNA templates are described initially, followed by the protocol for the transcription of mRNA from DNA templates and finally the method to perform eukaryotic cell-free reactions for the purpose of drug target production is depicted.

2.1.1. Design and Generation of DNA Templates

DNA templates form the basis for eukaryotic cell-free synthesis. Special sequence characteristics enable the use of DNA templates as expression sequences for cell-free protein synthesis. 5′ and 3′ UTRs need to be adapted for eukaryotic cell-free synthesis including T7 promotor and T7 terminator sequences for the transcription reaction and stem-loops for an increased stability of mRNA. As described in Figure 1, depending on the application, linear as well as circular DNA templates can be applied directly to cell-free reactions. Special regulatory sequences are required for wheat-germ-based cell-free synthesis. The next section deals with the design and generation of optimal DNA templates.

Plasmids

Plasmids required for linear DNA template generation and cell-free protein synthesis were obtained from GeneArt (ThermoFisher Scientific, Regensburg, Germany). This included the following plasmids:

- Human telomerase (hTERT): pMA-hTERT-His, (*Sf*21 and CHO cell-free synthesis), pMA-hTR (*Sf*21, CHO and wheat germ cell-free synthesis)
- WNT: pcDNA3.1(+)-WNT3a, pcDNA3.1(+)-WNT5a, pcDNA3.1(+)-WNT5b, pcDNA3.1(+)-WNT6 (linear DNA template generation and *Sf*21 cell-free synthesis)

Further plasmid templates were cloned at Fraunhofer IZI-BB. This comprises the following plasmids:

- pIX4.0-Luc (*Sf*21 and CHO cell-free synthesis)
- pIVEX1.3-hTERT

All plasmids harbor the previously described regulatory sequences.

Generation of Linear Expression DNA Templates

Linear WNT DNA templates were generated by Expression PCR (EasyXpress Linear Template Plus Kit Plus, Biotech-Rabbit GmbH) to exchange the native signal peptide by a melittin signal peptide (Mel) and thereby increasing translocation efficiencies as reported by Kubick et al. [25]. The first step of template generation included gene (forward) and plasmid specific (reverse) primers amplifying the desired linear DNA template based on pcDNA3.1(+)-WNT plasmids. For the attachment of 5'- and 3'- regulatory sequences, a second PCR step was performed using a melittin signal sequence primer (N-Mel) in combination with an antisense primer (C-0). Primers used for the amplification steps are listed in Table 1. The preparation of the PCR procedure was performed according to the manufacturer's instructions (EasyXpress Linear Template Plus Kit Plus, Biotech-Rabbit GmbH).

Mel-WNT3a harboring a C-terminal eYFP-tag (Mel-WNT3a-eYFP) was generated for the detection of cotranslational translocation of WNT-proteins into microsomal structures by confocal laser scanning microscopy. A two-step overlap extension (oe) PCR was performed for the amplification of WNT-eYFP fusion DNA templates. The first PCR step included a separate amplification of WNT-oe and oe-eYFP fragments. Gene-specific forward primer sequences and reverse primer sequences are listed in Table 2. During the second PCR step, both fragments were fused to each other. Furthermore, the 5'-end regulatory sequences and the melittin signal sequence were added using the adapter primer N-Mel and C-terminal antisense primer (C-SII). All PCR products were analyzed by 1% agarose gel electrophoresis and ethidium bromide staining. Primers were purchased from IBA (IBA GmbH, Göttingen, Germany).

Table 1. Primer sequences used for WNT DNA template generation including melittin signal peptide (Mel-WNT).

Primer	Primer Sequence 5' → 3'
Gene-specific forward (F) primers	
X-Mel-WNT3aOpt-F	TAC ATT TCT TAC ATC TAT GCG GAC TCC TAC CCC ATC TGG TGG TC
X-Mel-preWNT5aOpt-F	TAC ATT TCT TAC ATC TAT GCG GAC TTC GCT CAG GTC GTG ATC GAG GC
X-Mel-matWNT5aOpt-F	TAC ATT TCT TAC ATC TAT GCG GAC ATC ATC GGT GCT CAG CCC CTG T
X-Mel-WNT5bOpt-F	TAC ATT TCT TAC ATC TAT GCG GAC CAG CTG CTG ACC GAC GCT AAC TC
X-Mel-WNT6Opt-F	TAC ATT TCT TAC ATC TAT GCG GAC CTG TGG TGG GCT GTG GGT TC
Plasmid specific reverse (R) primer	
pcDNA3-R	CAA AAA ACC CCT CAA GAC CCG TTT AGA GGC CCC AAG GGG AGA AGG CAC AGT CGA GGC TG

Table 1. *Cont.*

Primer	Primer Sequence 5′ → 3′
	Adapter primers
N-Mel	ATGATATCTCGAGCGGCCGCTAGCTAATACGACTCACTATAGGGAGACCACAACGGT TTCCCTCTAGAAATAATTTTGTTTAACTTTAAGAAGGAGATAAACAATGAAATTCTTA GTCAACGTTGCCCTTGTTTTTATGGTCGTATACATTTCTTACATCTATGCGGAC
C-0	TAATAACTAACTAACCAAGATCTGTACCCCTTGGGGCCTCTAAACGGGTCTTGAGGG GTTTTTTGGATCCGAATTCACCGGTGAT

Table 2. Primer sequences used for the generation of eYFP c-terminally fused to Mel-WNT3a (Mel-WNT3a-eYFP).

Primer	Primer Sequence 5′ → 3′
	Gene-specific reverse (R) primer
X-Mel-WNT3aOpt-oeXFP-R	CTT GCT CAC CTC TAG ACA GGG CAC CTT TCC AGC G
	eYFP primer
X-eXFP-SR	TTG CGG ATG AGA CCA GGC AGA CTT GTA CAG CTC GTC CAT GC
oe-eXFP-F	TGT CTA GAG GTG AGC AAG GGC GA
	Adapter primer
C-SII	TGT CTA GAG GTG AGC AAG GGC GA

2.1.2. Transcription Reaction

The required mRNA for cell-free protein synthesis was prepared using an in vitro transcription approach. This technique is based on the application of a T7 RNA polymerase, which recognizes the T7 promotor of generated DNA templates for initiating mRNA transcription. In vitro transcription was carried out as described by Orth et al. [26] and transcription reaction was incubated for 2 h at 37 °C. Transcribed mRNA was purified prior to the application in cell-free translation reactions. Therefore, an intermediate gel filtration step is performed using a DyeEx 2.0 spin column (Qiagen, Hilden, Germany). The concentration of mRNA was analyzed using a Nanodrop 2000c spectrophotometer (ThermoFisher Scientific). Typically 1.0–1.3 nM of mRNA of different target proteins was added to translation reactions.

2.1.3. Preparation of Eukaryotic Lysates

Three different eukaryotic lysates were applied to perform the cell-free synthesis of target proteins. Eukaryotic lysates were based on CHO cells, *Sf*21 cells and wheat germ and can be separated into microsome including (CHO and *Sf*21) and microsome absent (wheat germ) lysates. CHO and *Sf*21 lysates for eukaryotic cell-free synthesis were prepared at Fraunhofer IZI-BB. First, the appropriate cell lines (CHO-K1 (ECACC 85051005) and *Sf*21 (DSM ACC 119)) were cultured in a batch mode fermentation (equipment Sartorius AG). Process parameters (pO_2, pH, temperature) were monitored and regulated to obtain a highly controlled lysate production process. Cells were harvested in the exponential growth phase when reaching a cell density between 4–6×10^6 cells/mL using centrifugation. The cell pellet was washed with HEPES based buffer (100 mM HEPES-KOH, 100 mM KoAc, 4 mM Dithiothreitol (DTT)) to remove media components. A mild cell disruption procedure was performed using a 20-gauge needle and raw lysate was separated from cell debris and nuclei by centrifugation ($10,000 \times g$ for 15 min at 4 °C). The raw lysate was further desalted and concentrated by size exclusion chromatography using a Sephadex G-25 column (GE Healthcare, Munich, Germany). Subsequently, A260 values were determined by Nanodrop 2000 (ThermoFisher Scientific) measurement and fractions with A260 values were pooled. Endogenous mRNA was digested by treating the pooled lysate with micrococcal S7 nuclease (10 U/µL, Roche, Penzberg, Germany). Nuclease activity was inhibited by the supplementation of 6.7 mM EGTA (ethylene glycol-bis(β-aminoethyl ether)-N,N,N′,N′-tetraacetic acid) (f. c.). Finally, CHO- and *Sf*21-lysates were aliquoted, shock-frozen in liquid nitrogen and stored at −80 °C. Wheat Germ lysate is commercially available in the form of RTS100 Wheat Germ CECF Kit (Biotechrabbit, Berlin, Germany).

2.1.4. Eukaryotic Cell-Free Protein Synthesis

Individual eukaryotic cell-free protein synthesis systems differ in their composition and reaction conditions. *Sf*21- and CHO-based cell-free reactions have similar characteristics and were performed in a batch-formatted reaction mode. A 25 μL standard translation reaction of a *Sf*21 and CHO based cell-free synthesis was composed of 6 μL purified mRNA, 40% lysate, canonical amino acids (200 μM each), ATP (1.75 mM), GTP (0.45 mM) and ^{14}C-labeled leucine (200 dpm/pmol) for the detection of de novo synthesized proteins. For the functional analysis of WNT proteins (β-catenin accumulation assay), proteins were synthesized in the absence of ^{14}C-leucine. Protein translation reactions based on *Sf*21 lysates were incubated for 90 min at 27 °C, 600 rpm using a thermomixer (Eppendorf, Hamburg, Germany). Translation reactions based on CHO cell lysates were performed at 30 °C and 120 min with gentle shaking at 600 rpm. If required translation mixture (TM) of both cell-free reactions were further fractionated for analysis of protein translocation. The fractionation was realized by centrifugation at 16,000× *g* for 10 min at 4 °C in order to separate the ER-derived microsomal fraction (MF) of the cell lysate from the supernatant (S). The microsomal fraction was resuspended in PBS buffer without calcium and magnesium ions for further analysis.

To decrease the phosphorylation of eukaryotic translation factor eIF2α and thereby improving the capacity of cap-dependent translation initiation, the influence of the small component C38 (GSK2606414, GlaxoSmithKline, Dresden, Germany), a specific PERK inhibitor, was analyzed. Inhibitor containing lysate was preincubated for 10 min at RT before the cell-free reaction was started. To analyze the effect on the protein synthesis rate, the pIX4.0-Luc plasmid was added to the reaction and the yield of active luciferase was determined.

Synthesis in wheat germ lysate was performed in a transcription/translation coupled dialysis system using the RTS100 Wheat Germ CECF Kit (Biotechrabbit) for 24 h at 24 °C according to the manufacturer's instructions. Again, ^{14}C-leucine (2.47 dpm/pmol) was added to the reaction mixture to determine the size, yield and integrity of the de novo synthesized protein.

To obtain functional active hTERT enzyme, the assembly with a typical RNA component of telomerase is required [27]. The assembly was realized in two different ways: RNA was generated in a previous transcription reaction by using T7 RNA–polymerase followed by RNA purification using DyeEx spin columns (Qiagen). The freshly synthesized hTR RNA was directly added to the cell-free reaction (CHO and *Sf*21 cell-free reactions). Alternatively, a plasmid harboring the nucleotide sequence encoding the RNA component of human telomerase under control of the T7-promoter was added directly to the translation reaction (coupled reaction; wheat germ; final concentration of hTR plasmid: 20 ng/μL).

2.1.5. Detection of eIf2 Phosphorylation

Western Blots analysis and ELISA were performed to monitor the presence of phosphorylated eukaryotic initiation factor eIF2 in eukaryotic cell-free systems. Phosphorylation of subunit α of eIF2 usually inhibits cap-dependent translation initiation in eukaryotic cells. This might also influence the translational activity of eukaryotic cell lysates. To enable the detection of eIF2α-P with western blot, the translation mixture of a CHO batch synthesis was acetone precipitated and present proteins were separated by SDS-PAGE according to the standard protocol (Section 2.2.3). After finishing the SDS-PAGE standard separation protocol, the gel was placed onto the "IBlot Gel Transfer Device" (ThermoFisher Scientific) and blotting was performed according to the manufacturer's protocol. After finishing the blotting procedure, PVDF membrane containing desired proteins was washed three times in TBS for 10 min. Next, the membrane was blocked using 2% BSA in TBS/T at RT for 2 h or at 4 °C overnight followed by three times washing with TBS/T for 10 min. Subsequently, the membrane was incubated with gentle agitation for 3 h with the desired rabbit anti-eIF2α-P primary antibody (Cell Signaling Technology, Frankfurt, Germany) prediluted in a 2% BSA-TBS/T solution (1:1000). The previously described washing procedure was repeated and Anti-rabbit IgG, HRP-linked antibody (1:2000) (Cell Signaling Technology) was applied to the membrane. Conjugated horseradish peroxidase (HRP) enabled the detection of target protein bands. HRP substrate Amersham ECL select western blot detection reagent (Promega, Mannheim, Germany) was added to the membrane and incubated for 5 min. To visualize the specific protein bands chemiluminescence signal was analyzed using a Typhoon Trio + variable mode imager (GE Healthcare). A detection of eIF2α-P by ELISA was performed using a PathScan® Phospho-eIF2α (Ser51) Sandwich ELISA according to the manufacturer's protocol.

2.2. Qualitative and Quantitative Analysis of Cell-Free Synthesized Proteins

Methods and materials used for the analysis of cell-free synthesized proteins are described. In this section quantification, molecular size detection and functional analysis of the presented, cell-free synthesized proteins are introduced in detail.

2.2.1. Luciferase Assay

For the optimization of translational activity and stability of eukaryotic cell lysates, a firefly luciferase was used as a model protein. Functional luciferase was quantified using a standard luciferase assay (Promega). Therefore, 5 µL of the cell-free translation mixture was applied to a white Nunc 96-well microtiter plate and mixed with 50 µL luciferase assay reagent. The luciferase assay was performed using a TriStar LB 941 multimode reader (Berthold Technologies, Bad Wildbach, Germany). The amount of functional protein was calculated based on the obtained relative light units (RLU), according to the following equation (calibration curve):

$$\text{Active luciferase} = 7 \times 10^{-5} \times \text{RLU} \tag{1}$$

The amount of active luciferase was calculated from a triplet of three independent experiments (n = 3). Mean values and standard deviations were estimated as described in the TCA precipitation and scintillation measurement section.

2.2.2. Quantification of Cell-Free Synthesized Protein Yield

Based on the incorporation of ^{14}C-leucine in cell-free synthesized proteins, the respective protein yield can be estimated by scintillation measurement. Therefore, 5 µL aliquots of each translation mixture were mixed with 3 mL of a 10% (v/v) TCA-2% (v/v) casein hydrolysate solution (Carl Roth, Karlsruh, Germany) in a glass tube and incubated at 80 °C for 15 min. Afterwards, samples were chilled on ice for 30 min and retained on the surface of glass fiber filter papers (MN GF-3, Machery-Nagel, Düren, Germany) using a vacuum filtration system (Hoefer, Kleinblittersdorf, Germany). Filters were washed twice with 5% TCA and dried with acetone. Dried filters were placed into a scintillation vial (Zinsser analytic, Eschborn, Germany), 3 mL of scintillation cocktail (Quicksafe A, Zinsser analytic) was added and vials were agitated on an orbital shaker for at least 1 h. The scintillation signal was determined using the LS6500 Multi-Purpose scintillation counter (PerkinElmer, Berlin, Germany). The protein concentration was identified based on the obtained scintillation counts and protein specific parameters including molecular mass and amount of leucine. Error bars calculated for the protein yield show the individual standard deviation.

2.2.3. SDS-PAGE and Autoradiography

The molecular size of radiolabeled, cell-free synthesized protein was analyzed using SDS-PAGE followed by autoradiography. First, 5 µL of the respective fraction of a cell-free synthesis reaction including the radiolabeled target protein was subjected to ice-cold acetone. Precipitated protein was separated by centrifugation (16,000× *g*, 4 °C, 10 min) and protein pellet was dried for at least 30 min at 45 °C. The dried protein pellet was dissolved in LDS sample buffer (ThermoFisher Scientific), heated at 70 °C for 10 min and loaded on a precasted NuPAGE 10% Bis-Tris gel (ThermoFisher Scientific). The gel was run at 185 V for 35 min according to the manufacturer's protocol. Subsequently, gels were dried at 70 °C (Unigeldryer 3545D, Uniequip, Planegg, Germany), placed on a phosphor screen and radioactively labeled proteins were visualized using a Typhoon Trio + variable mode imager (GE Healthcare).

2.2.4. β-Catenin Accumulation Assay

The functional activity of cell-free synthesized WNT protein (Mel-WNT3a) is analyzed based on canonical WNT signaling. In the applied assay, accumulation of β-catenin was analyzed using a western blot detection approach (described in Figure 3D). For this, confluently grown HeLa cells were stimulated with microsomes dissolved in cell culture media (DMEM, Merck Millipore, Darmstadt, Germany) with 1% Chaps (Sigma Aldrich, Taufkirchen, Germany) including cell-free synthesized Mel-WNT3a (50 ng/μL and 100 ng/μL). Recombinant WNT3a (RnD systems) and LiCl (50 mM, Sigma Aldrich) served as positive controls. To analyse background β-catenin expression, untreated HeLa cells (UTC) and stimulation with dissolved microsomes without cell-free synthesized WNT protein (NTC, volume equivalent to 100 ng/μL WNT3a volume) were used for the assay. After treatment of cells (3 h, 37 °C) with samples and control reagents, HeLa cells were lysed using RIPA-Buffer (50 mM Tris-HCl, 150 mM NaCl, 1% IGEPAL, 0.5% Sodium deoxycholate, 0.1% SDS, complete protease inhibitor cocktail tablets (Roche)). Cell lysates were further homogenized and centrifuged at 4000× *g* for 10 min at 4 °C. Supernatants were applied to a western blot. To enable comparability of the samples, total protein yield was determined using the Pierce BCA Protein Assay Kit (ThermoFisher Scientific) according to the manufacturer´s protocol. 20 μg of the total protein was loaded on a SDS-PAGE for further Western blot analysis (description of the method is in Section 2.1.5). Detection of β-catenin was realized using a primary rabbit anti-β-catenin antibody (1:1000) (Cell Signaling) and a secondary anti-rabbit HRP (Horseradish peroxidase) conjugated antibody (1:2000) (Cell Signaling Technologies). Detection of a β-actin was performed using a rabbit anti-β-actin antibody (1:1000) (Cell Signaling Technologies). The housekeeping protein β-actin served as an internal control and for normalization of image analysis.

2.2.5. Analysis of Protein Translocation Using Confocal Laser Scanning Microscopy

In selected eukaryotic cell-free systems (*Sf*21 and CHO based system), microsomal structures are presently derived from the endoplasmic reticulum. These microsomes enable a direct translocation of secreted and membrane-embedded proteins followed by ER-based posttranslational modifications. In this study, the translocation of secreted eYFP fusion proteins (Mel-WNT-eYFP) into ER-derived microsomal structures was analyzed by confocal laser scanning microscopy (CLSM) using an LSM Meta 510 microscope (Zeiss, Oberkochen, Germany). The preparation of samples included a dilution of TM (5 μL) with PBS (15 μL) and a transfer of 20 μL diluted sample to μ-slides (Ibidi, Planegg, Germany). Samples were excited at 488 nm using an argon laser. The emission was detected using a band pass filter in the range of 500–550 nm.

2.2.6. Telomerase Activity Assay

The functionality of telomerase was evaluated using the TeloTAGGG Telomerase PCR ELISA PLUS Kit (Roche). The general principle of this assay is illustrated in Figure 4B. In detail, 1 μL aliquots of the translation mixture were used for the amplification reaction where active telomerase adds telomeric repeats (TTAGGG) to the 3′-end of a biotin-labeled primer. The resulting products, harboring the telomere-specific repeats, were amplified again. In the next step, a 2.5 μL aliquot of the product was denatured and hybridized to a digoxigenin-(DIG)-labeled probe that recognizes telomeric repeats. The product consisting of telomeric repeats, the DIG-probe and the biotin-labeled primer were immobilized to a streptavidin-coated microplate and detected with a peroxidase-conjugated antibody against digoxigenin (anti-DIG-POD). By adding tetramethylbenzidine (TMB) substrate a colored complex was formed displaying an absorbance which could be measured at 450 nm with a reference wavelength of 690 nm. The relationship between the absorption of an internal standard and the sample gives the relative telomerase activity. Activity assays were performed in two independent experiments and standard deviations were calculated.

3. Results

In this section, the performance of eukaryotic cell-free systems for the production of druggable targets is exemplarily shown. The chapter starts with general strategies to obtain an optimal eukaryotic cell-free system for the synthesis of mammalian proteins. This is followed by two examples of drug target proteins, which are produced and functionally characterized in eukaryotic cell-free protein synthesis systems. In this case, secreted WNT-proteins and the cytosolically produced hTERT enzyme is chosen to show the potential of the system for future protein characterizations and drug development.

3.1. Strategies to Optimize Protein Production in Eukaryotic Cell-Free Systems

Protein translation, stability and preservation require certain conditions, which need to be realized in cell-free protein synthesis systems. In this study, eukaryotic cell-free protein synthesis systems are applied to produce drug target proteins. There are several weak points which can be addressed for the improvement of the eukaryotic cell-free systems. A general overview of the optimization strategies is pointed out in Figure 2A. In general, three main topics for the optimization of the systems are available. First, translation factors need to be present in a highly active form in the eukaryotic lysate. Cell cultivation and lysate preparation can influence the activity of translation factors due to nutrient limitations, oxidative stress and initiation of unfolded protein response. Partial inactivation of translation factors limits the production rate of eukaryotic cell-free protein synthesis systems. In addition, proteolytic enzymes in eukaryotic lysate can influence the stability of target proteins and translation associated factors. A general and targeted inhibition of proteases can improve cell-free systems. An example of the effective use of protease inhibitors is depicted by the application of caspase inhibitors in eukaryotic continuous exchange cell-free systems [22,28]. Apart from the lysate itself, general reaction conditions can be improved by adjusting energy components, buffer conditions and salt concentrations.

In Figure 2, the topic of translation initiation factors is addressed in the context of cap-dependent translation initiation in eukaryotic cell-free systems. Eukaryotic translation initiation is a complex process that requires the presence of numerous translation initiation factors in an active state. Inactivation of translation is a regulative tool in living cells to react to stress responses. One of the main initiation factors in terms of regulatory effects is the eukaryotic initiation factor 2 (eIF2). eIF2 can be subdivided into three subunits (α, β, γ), while phosphorylation of subunit α leads to factor inactivation and thereby inhibition of cap-dependent translation initiation. In Figure 2B–D activity of translation factor eIF2 was analyzed and factor phosphorylation was investigated in the presence of the specific phosphorylation inhibitor small component C38 (GSK2606414, GlaxoSmithKline). C38 specifically inhibits the protein kinase R (PKR)-like endoplasmic reticulum kinase PERK, one of the kinases responsible for eIF2α phosphorylation. For the experimental setup, cell-free reactions based on CHO cell lysate were used. eIF2α-P specific ELISA (Figure 2B) and western blot analysis (Figure 2C) were performed using untreated cell-free reactions (UTC), C38 treated cell-free reactions and DMSO treated cell-free reactions. DMSO, the solvent for C38, serves as a background control for eIF2α-P. ELISA results displayed an absorbance of 0.1 units in the UTC and the DMSO treated sample (Figure 2B). A decrease in absorbance was detected in the C38 treated sample (0.04 units). Similar results were obtained in western blot analysis (Figure 2C). Western blot analysis was performed using cell-free reaction samples taken at different time points of cell-free reaction (0 min, 30 min, 120 min). After 30 min and 120 min reduced intensities of eIF2α-P samples were detected after treatment with C38 in comparison to UTC and DMSO samples. The intensity of phosphorylation increases over time in UTC and DMSO samples. The result of both analyses depicts how the treatment of eukaryotic cell-free reaction with C38 led to a decrease in eIF2 phosphorylation. The following results show the influence of the inhibitor on the protein production rates (Figure 2D). Model protein Luciferase was chosen to evaluate the productivity of the system in the presence of C38. The yield of active Luciferase was evaluated using

Luciferase detection reagent and chemiluminescence analysis. An NTC consisting of eukaryotic cell-free reaction without synthesized protein was prepared as a background control for the assay. A significant increase in protein yield was detected by treatment with C38. While UTC and DMSO treated samples show low protein yields in the range of 0.05 µg/mL–0.1 µg/mL an increase up to 9.5 µg/mL was detected using 4.5 mM of C38. A saturation of protein yield was reached using 4.5 mM C38. These results underline the beneficial effect of C38 on the translational activity of eukaryotic cell lysates.

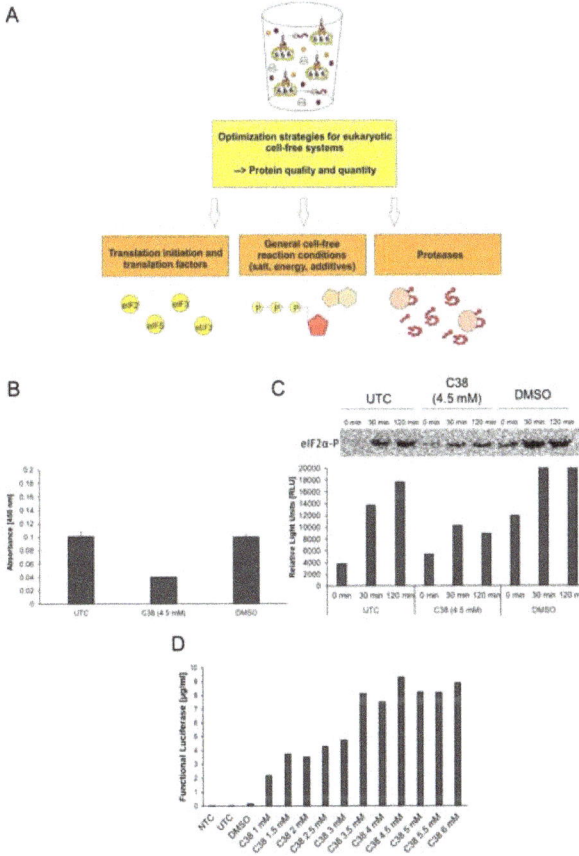

Figure 2. Optimization strategies for eukaryotic cell-free systems. (**A**) Schematic overview of strategies for improvement of eukaryotic cell-free systems concerning protein quality and quantity. (**B**) PathScan® Phospho-eIF2α (Ser51) Sandwich ELISA (Cell Signaling) for the detection of eIF2α-P in CHO cell-free reactions. Untreated control (UTC), DMSO treated samples (DMSO) and C38 treated samples were analyzed for the presence of eIF2α-P. ELISA was performed according to the manufacturer's protocol. (**C**) Western Blot analysis of UTC, DMSO treated samples (DMSO) and C38 treated samples using a primary anti-eIF2α-P antibody (Cell Signaling) and a secondary anti-rabbit-HRP antibody. Western blot signal was detected using ECL reagent (Promega) and a Typhoon Trio Plus Imager (GE Healthcare). Image analysis was performed using ImageQuant TL software (GE Healthcare). (**D**) Eukaryotic cell-free reaction in the presence of C38 PERK inhibitor. CHO lysate based cell-free reaction was performed using pIX4.0-Luc plasmid (Luciferase) in the presence of various concentrations (1 mM up to 6 mM) of small component C38. Production of cell-free synthesized Luciferase was analyzed using a standard Luciferase assay (Promega).

3.2. Synthesis of WNT Proteins in Sf21 Cell-Free Systems

WNT proteins belong to a family of signaling proteins which are essential for many biological processes. This highly conserved family of signaling molecules [25] is involved in body axis formation, embryonic growth, cell differentiation and proliferation as well as tissue homeostasis [29,30]. WNT signal transduction is one of the most important pathways in human development and maintenance. Thus, dysfunction of WNT pathway regulation in human organisms plays an important role in the formation of well-known diseases [31] including various cancer types, especially colon and lung cancer, diabetes, kidney disorders and neurodegeneration [32–34]. WNT proteins are secreted proteins harboring several posttranslational modifications (glycosylation, lipidation). Such modifications are mandatory to maintain the activity of the proteins. Research in the field of WNT signal transduction is often limited by the inaccessibility of the produced bioactive WNT proteins [35,36]. Therefore novel production systems are required to produce WNT proteins to simplify future drug development. In this chapter, the production of WNT proteins in eukaryotic cell-free systems (*Sf*21 cell-free system) is described. In the first set of experiments, four candidates of WNT proteins (WNT3a, WNT5a, WNT5b, WNT6) were selected (Figure 3A). WNT proteins were cell-free synthesized in the presence of ^{14}C leucine to enable further quantification and estimation of molecular weight by autoradiography. Two variants of the selected WNT proteins were synthesized and analyzed by autoradiography: The first variant harbors the original human WNT sequence including the human signal peptide, the second variant contains a melittin signal peptide instead of the human signal peptide. Previous studies showed an efficient translocation of secreted proteins into microsomes of *Sf*21 cell-free systems using the melittin signal peptide [37]. The exchange of human signal peptide to melittin signal peptide was realized by Expression PCR. Subsequently, PCR products were directly applied to the cell-free reaction. Prior to analysis, cell-free TM containing WNT proteins was separated into a SN fraction and MF to analyze the translocation of proteins into ER-derived microsomes. The apparent molecular mass of cell-free synthesized WNT ligands (listed in Table 3) was identified by SDS-PAGE followed by the autoradiography of radiolabeled proteins. Different types (WNT3a, WNT5a, WNT5b, WNT6) and variants (+Hum/+Mel) of WNT proteins showed bands at the expected molecular mass in the range of 39 kDa to 43 kDa (Figure 3A). In the TM and MF fractions, additional protein bands were detected displaying a slightly higher molecular mass than the expected protein bands, thereby indicating posttranslational modifications like glycosylation. Glycosylation of WNT ligands was further confirmed by PNGaseF treatment of TM samples (Supplementary Figure S1). The endoglycosidase treatment led to a disappearance of higher molecular mass bands on the autoradiograph thereby indicating the presence of glyco-modifications. Supernatant fractions of WNT proteins only harbor a single protein band at the expected molecular mass. Moreover, WNT protein variants containing a melittin signal peptide showed an increase in the intensity of protein bands in the microsomal fraction compared to WNT proteins with a human signal peptide. Therefore, autoradiography data imply differences in protein translocation into microsomes. In this case, melittin signal peptide also showed a more efficient translocation into microsomes of *Sf*21 cell-free systems. The translocation of WNT proteins into ER-derived microsomes was further verified by CLSM (Figure 3B). For this, suitable DNA templates for Mel-WNT-eYFP fusion proteins (WNT3a, WNT5a, WNT5b and WNT6) were generated to enable the visualization of protein translocation into ER-based microsomes. After finishing an *Sf*21 based cell-free synthesis of WNT-eYFP variants, pre-diluted translation reaction was analyzed by CLSM. To verify background fluorescence of the cell-free translation mixture, a NTC was prepared without any cell-free produced protein. A strong localization of the fluorescence signal in vesicle-like structures was detected for all WNT variants but lacked in the NTC sample, as expected. These results give a hind for successful translocation of WNT proteins into ER-based microsomes.

Table 3. Overview of molecular mass of WNT proteins.

WNT Protein	Molecular Mass (kDa) (Human Signal Peptide)	Molecular Mass (kDa) (Melittin Signal Peptide)
WNT 3a	42.9	43.5
WNT 5a	42.3	41.4
WNT 5b	40.3	40.3
WNT 6	39.7	39.3

For further characterization of cell-free synthesized WNT proteins, Mel-WNT3a was exemplarily chosen to show the analysis of cell-free reaction and functional characterization. Figure 3C showed the obtained protein yields of Mel-WNT3a in *Sf*21 lysate based cell-free systems. After 120 min reaction time, the maximum protein yield of 23 µg/mL Mel-WNT3a was obtained. Longer reaction times led to a slight degradation of WNT protein. On the basis of this result, a functional characterization of Mel-WNT3a was performed. The general principle of the applied assay is based on the activation of the canonical WNT signaling pathway. During canonical WNT signaling, the WNT protein binds to the cell membrane receptors LRP5/6 and Frizzled and thereby activates intracellular signaling processes. This leads to the accumulation of β-catenin, a transcription regulator which actives WNT specific gene expression. The schematic protocol of the assay is illustrated in Figure 3D. The microsomal fraction harboring cell-free synthesized Mel-WNT3a is treated with 1% Chaps in cell culture medium to release translocated WNT proteins. The dissolved cell-free-WNT mixture (50 ng/µL and 100 ng/µL) is directly applied to HeLa cells and after 3 h of incubation, cells are lysed and accumulation of β-catenin is evaluated by western blot analysis. Apart from stimulation with cell-free produced WNT protein, control simulations were performed using cell-free reaction without synthesized protein (NTC), recombinant commercially available WNT protein (rec. WNT3a (50 ng/µL and 100 ng/µL)) and LiCl as a signaling stimulator (LiCl). UTC was taken as a background control. Results from the western blot analysis (Figure 3E) show the highest intensity of the β-catenin protein band by applying cell-free produced WNT-protein as a stimulant. Increased intensities were also detected for recombinant WNT and LiCl in comparison to the NTC and UTC sample. Image analysis where a β-actin housekeeping gene was used for normalization confirm these results.

The obtained results for the production of WNT proteins in *Sf*21 cell-free systems underline the potential for the fast production of complex and functional signaling proteins in eukaryotic cell-free systems.

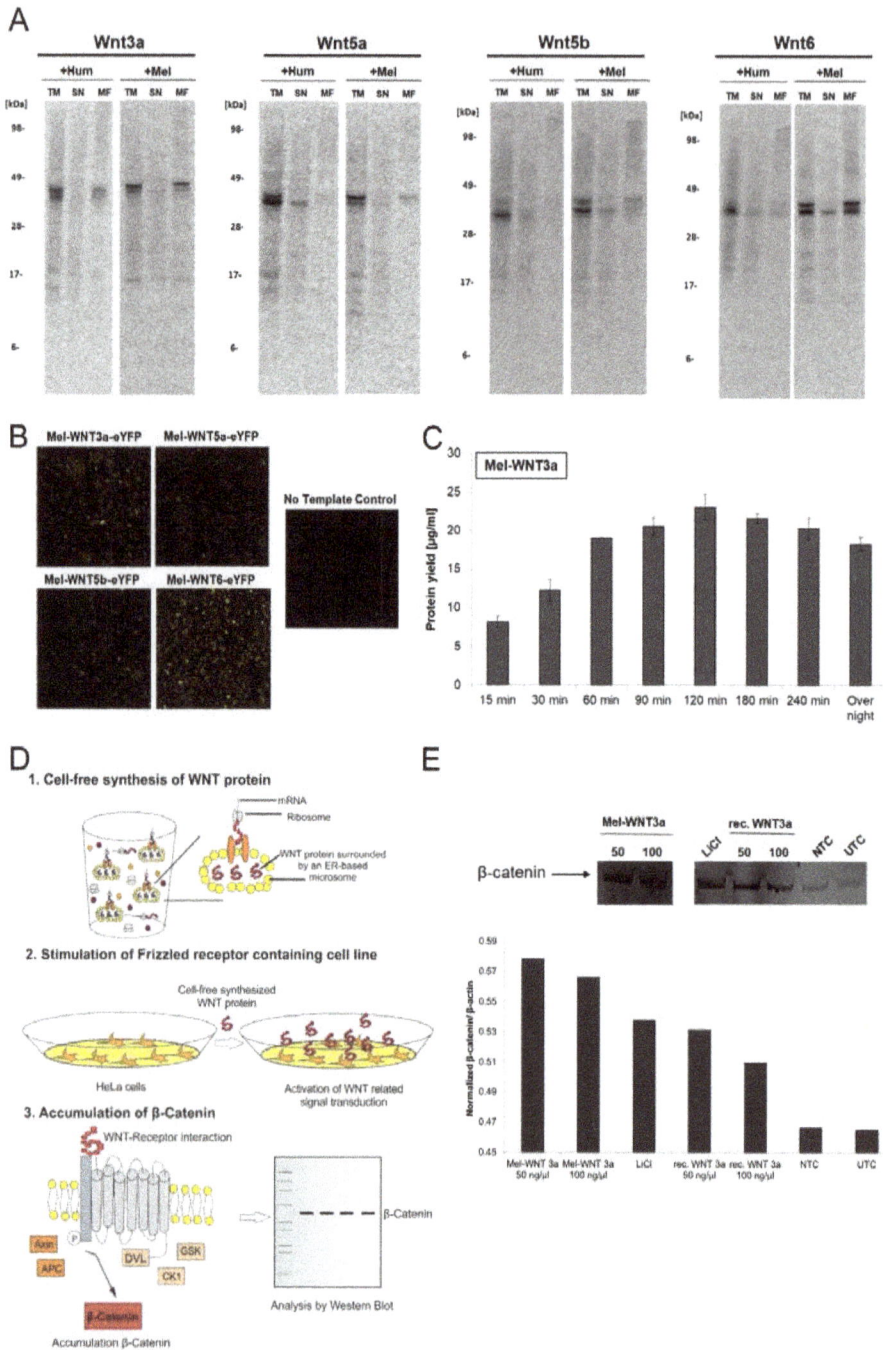

Figure 3. Cell-free synthesis of WNT proteins using *Sf*21 cell lysate. WNT proteins were synthesized in a batch-formatted *Sf*21 cell-free system. (**A**) Autoradiography of cell-free synthesized WNT proteins (WNT3a, WNT5a, WNT5b, WNT6). The qualitative analysis shows the expected molecular weight of

the produced proteins. Production of WNT types was carried out with a human signal peptide (on the basis of pcDNA3.1(+)-WNT) and with a melittin signal peptide (on the basis of Expression PCR product). Translation mixture (TM) of cell-free synthesized WNT proteins were separated by centrifugation into a supernatant fraction (SN) and microsomal fraction (MF). (**B**) Analysis of translocation of WNT-eYFP fusion proteins by confocal laser scanning microscopy. (**C**) Analysis of Mel-WNT3a protein yield in a time course experiment. Radiolabeled proteins were TCA precipitated, transferred to a filter paper and dissolved in scintillation cocktail. Protein yield was determined by scintillation measurement using an LS6500 multi-purpose scintillation counter (PerkinElmer). (**D**) Schematic overview of the assay for functional characterization of WNT proteins based on canonical WNT signaling. (**E**) Functional characterization of cell-free synthesized Mel-WNT3a (50 ng/µL and 100 ng/µL) using the previously illustrated assay. Western blot analysis of accumulated β-catenin using primary anti-β-catenin and secondary anti-rabbit-HRP antibody (Cell Signaling). Protein band was detected using an ECL reagent (Promega) and a Typhoon Trio Plus Imager (GE Healthcare). Image analysis was performed with a housekeeping gene (β-actin) for normalization using ImageQuant TL software (GE Healthcare). NTC: No template control (cell-free reaction without WNT DNA template).

3.3. Eukaryotic Cell-Free Systems for the Production of Human Telomerase

Telomerase is a ribonucleoprotein which adds hexanucleotides like TTAGGG in vertebrates to the ends of chromosomes and thereby protects the telomeres against aging caused by shortening during cell division. The reverse transcriptase subunit (hTERT) is inactive in most somatic cells but active in nearly all cancer cells showing an unlimited proliferation. For these reasons, it is essential to produce active recombinant telomerase in sufficient amounts for subsequent pharmacological characterization. In vivo expressed hTERT is usually insoluble due to incorrect protein folding. Furthermore, hTERT requires the assembly with a telomerase intrinsic RNA (hTR) to form a completely active ribonucleoprotein. In this chapter, the production of hTERT in cell-free protein synthesis systems was evaluated. Apart from the previously described *Sf*21 and CHO lysate based cell-free system, a third eukaryotic cell-free system based on wheat germ extract was evaluated for the synthesis of hTERT. Protein production using wheat germ lysate differs significantly from previously used systems due to the reaction mode (dialysis mode). The production efficiency and the functional activity of hTERT proteins produced in eukaryotic lysates were compared (Figure 4). In all three systems the presence of a hTERT protein band at an appropriate molecular mass of 127 kDa was detected (Figure 4A). The intensity of the protein band differs in all three systems, which is equivalent to the obtained protein yields (Figure 4B) where large differences were detected (CHO: 3 µg/mL, *Sf*21: 28 µg/mL, wheat germ: 1.5 mg/mL). The highest protein concentration was obtained in the wheat germ dialysis systems with up to 1.5 mg/mL.

As mentioned above, one of the fundamental advantages of cell-free systems is their open nature. Therefore additional components can be easily added to the reaction to improve the quality of the produced protein. In this case, the telomerase intrinsic RNA (hTR) was added in the form of a previously transcribed RNA to CHO and *Sf*21 based batch-systems. Alternatively, hTR encoding plasmid was added to wheat-germ-based dialysis systems. The general principle of the applied TeloTAGGG Telomerase PCR ELISAPLUS Kit (Roche) is illustrated in Figure 4C. Telomerase activity referring to the synthesis of the enzyme in the presence and absence of hTR is given in Figure 4D. 1 µL of each cell-free translation mixture was directly used to perform teloTAGGG assay, hTERT produced in the presence of hTR in *Sf*21 and wheat germ lysates displays functional activity, while the total activity of wheat germ based synthesis is approximately increased by 25% in comparison to *Sf*21 extract based production. With a focus on protein yield, hTERT produced in an *Sf*21 cell-free system showed a significant increased relative activity in comparison to wheat germ hTERT. In contrast to the previously mentioned platforms, hTERT synthesized in CHO cell-free systems showed no activity, but a high background activity could be detected in the negative control (cell-free reaction without synthesized hTERT).

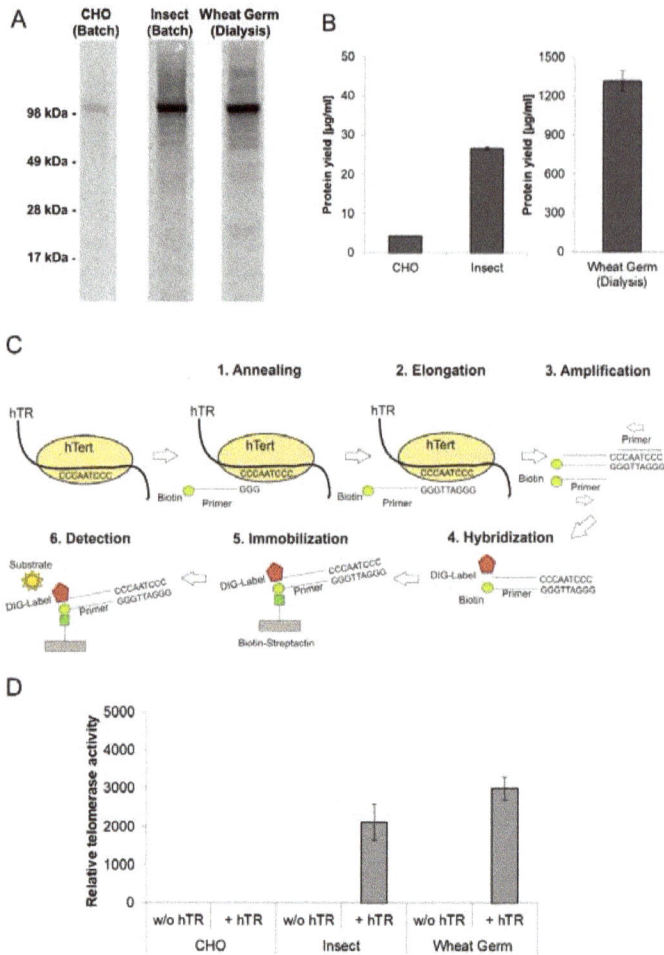

Figure 4. Quantitative and qualitative characterization of cell-free synthesized human telomerase (hTERT). Synthesis of hTERT in eukaryotic cell-free systems in batch (CHO and Sf21) and dialysis (wheat germ) mode. Protein was labeled with ^{14}C-leucine for further analysis. (**A**) A 5 µL aliquot of translation reaction mixture was precipitated with acetone and the resulting protein pellets were resolved in sample buffer. Samples were electrophoretically separated on a 10% SDS-PAGE gel followed by autoradiography. (**B**) Determination of hTERT protein yield by scintillation measurement. 5 µL of translation mixture was precipitated using TCA and precipitated protein was soaked on a filter sheet. The filter sheet was dissolved in scintillation liquid and analyzed using an LS6500 multi-purpose scintillation counter (PerkinElmer). (**C**) The general principle of hTERT activity assay (TeloTAGGG Telomerase PCR ELISA PLUS Kit (Roche)). (**D**) Functional characterization of cell-free synthesized hTERT using TeloTAGGG Telomerase PCR ELISAPLUS Kit. Synthesis of hTERT was carried out in the absence and presence of telomerase-specific RNA (hTR). According to the manufacturer's protocol (TeloTAGGG Telomerase PCR ELISA PLUS, Roche), samples are considered as telomerase-positive if the difference between the sample and the negative control is higher than the twofold value of the negative control. Samples, where the difference is lower than the twofold value of the negative control, were considered as telomerase-negative and set to zero relative telomerase activity.

4. Discussion

Over the recent years, cell-free protein synthesis was developed to a sophisticated tool for a broad range of applications in research and in the industrial context. Starting in the early 1960s with the first *E.coli* based cell-free system from Nirenberg et al. [11] a huge development of cell-free systems took place until now. Versatile platforms are available based on eukaryotic and prokaryotic cell lysates addressing a broad variety of applications ranging from biomedical areas to biofuel research [10]. A growing demand for proteins is noticeable in the biopharmaceutical industry, which is required on the one hand for the production of novel pharmaceuticals and on the other hand serve as target proteins for the development of drugs. Most of these proteins originate from a human organism or harbor human-like modifications, therefore special needs for the utilized production system are required. Eukaryotic cell-free protein synthesis systems mostly fulfill these requirements by enabling ER based posttranslational modifications and these systems harbor the set of cofactors and chaperones necessary for the correct folding and assembly of human proteins [18]. For a long time, eukaryotic cell-free systems were costly and inefficient, thereby drew limited attention for market applications. However, in recent years, the technology has been improved drastically [20,22,28,38] and achieved a special interest in terms of commercialization. In this article, the production of druggable, mammalian target proteins in eukaryotic cell-free protein synthesis systems is highlighted. Improvement of target protein quality and translation efficiency can be addressed by different strategies starting with the direct improvement of cell lysate by adaptation of cultivation strategies, activation and enrichment of translation-relevant factors and inhibition of protease activities. Various strategies for process optimization are reported in different cell-free systems. For the topic of cell-free reaction conditions, Caschera and Noireaux [39] underlined the importance of preparation of amino acid mixture and relevance of pH and concentration to obtain maximum protein yields in cell-free systems. Apart from this, several articles showed the relevance of precisely adjusted ion concentrations for optimal translation efficiencies [40,41]. Beyond general reaction conditions, optimizations in the area of translation factors are reported for HeLa based cell-free systems. Mikami and colleagues supplemented the mammalian cell-free system with recombinant translation factors eIF4E, eIF2 and eIFB, resulting in an improved protein translation rate [42]. Another approach was the elimination of ribosome inactivation factors in *Bacillus subtilis* and *S. cerevisiae* by genomic disruption which robustly improves the activity of cell-free systems [43]. In this article, the CHO based eukaryotic translation systems were improved by addressing the activation of translation factor eIF2 using the small component inhibitor C38. C38 inhibits the eIF2α specific kinase PERK [44], which phosphorylates the protein and thereby led to inactivation of cap-dependent translation initiation. The results showed that phosphorylation of eIF2α increased during the cell-free reaction, which might be due to ER stress and the presence of unfolded proteins. ER-stress is one of the most relevant factors for activation of ER-based PERK [45] which might be activated by mechanical disruption procedures during lysate preparation. The application of C38 is an efficient method to improve cap-dependent translation initiation in eukaryotic cell-free systems harboring microsomes. The described approach enables a fast and easy improvement of cap-dependent translation initiation independent of the type of eukaryotic cell lysate. By supplementation of C38, a significant improvement of protein yield is obtained. Additional future approaches in the area of eIF2α activation might be performed with a special focus on cost efficiency by directly addressing the cellular signaling pathways. It is conceivable, for instance, to directly engineer the stress response signaling pathway or diminish the expression of PERK in eukaryotic cells, thereby improving the cell-free platform. Similar approaches are already published for *E.coli* based cell-free systems where release factor 1 is depleted to improve incorporation of non-canonical amino acids using amber stop codon technology [46]. In this context, cell engineering strategies are an additional method for the future improvement of eukaryotic cell-free systems.

Optimized cell-free systems can be directly used for the production of druggable target proteins. In this study, WNT proteins and hTERT were exemplarily chosen to be synthesized and functionally characterized in eukaryotic cell-free systems. In conventionally used protein production platforms both proteins can hardly be expressed due to their "difficult-to-express" nature [47–49]. The attention for research correlating with WNT signaling is steadily rising mainly because of its fundamental role in the development of numerous diseases [50]. Several diseases are correlated with specific mutations in different components of the WNT signaling pathway [51], while targeted drugs need to be developed to address the mutated proteins. In this study, we have shown the production of different WNT protein ligands in an *Sf*21 based cell-free system. Additionally, we could prove the translocation of the proteins into ER-based microsomes, a requirement for correct posttranslational modifications and correct folding of WNT proteins. The eukaryotic cell-free system showed translocation efficiencies between 20%–40% of total translated protein. Former experiments underline that limitations of translocation are not linked to a limited volume of microsomes [52]. Furthermore, a prolongation of translation reaction by shifting the reaction format to a two-chamber dialysis system led to an increase of translocated membrane protein EGFR up to 60%–70% [22]. During dialysis reaction translation speed is continuously slowed down by gradually decreasing concentrations of energy components which in turn might support a proper ribosomal arrest for assembly of the translocation machinery. An additional factor might be the limited availability of translocation correlated factors like signal recognition particles (SRP) and signal recognition receptors. For cell-based protein production, it is reported that overexpression of SRP led to an increase in antibody secretion [53]. Similar approaches are conceivable for future system improvements. It is noticeable that translocation of WNT proteins into microsomes of *Sf*21 lysate is more efficient using a melittin signal peptide instead of the native human signal peptide of WNT proteins. The compatibility of melittin signal peptide and *Sf*21 cell-free systems was already reported in former studies [54]. The results showed that the eukaryotic cell-free system can not only be used for the production of defined proteins but also for comparison of their signal peptides. This technique paves the way for a future application to evaluate and optimize signal peptide sequences in a fast and high throughput manner prior to protein expression in cell-based systems. In addition to signal sequence optimization, Mel-WNT3a was functionally characterized showing increased biological activity in comparison to commercially available WNT proteins. In eukaryotic cell-free systems, WNT proteins are translocated into ER-derived microsomes and further post-translationally modified by palmitoylation and glycosylation. In eukaryotic cell-free systems, ER based posttranslational modifications are possible due to the presence of microsomes. These microsomes harbor the set of glycosyltransferases present in the ER leading to the formation of core glycosylation, which is identical in various eukaryotic cell-free systems. The presence of glycan moieties on WNT ligands might be beneficial for the correct folding of the protein. Palmitoylation might lead to an anchoring of WNT proteins to microsomal membranes thereby improving their biological activity. Former studies showed an improvement of WNT activity by tethering WNT3a to a liposomal membrane [55]. Water soluble and biologically active forms of WNT3a can also be obtained in the presence of albumin [47]. Eukaryotic cell-free systems with endogenous microsomes proved to be the most efficient platform for the production of biologically active WNT proteins. These results open up new opportunities for the detailed analysis of signaling processes and the characterization of druggable target proteins. Signal transduction recreated in eukaryotic cell-free systems bears an enormous potential for engineering and analysis of cellular pathways. By providing the opportunity to synthesize whole signal transduction pathways in a eukaryotic cell-free system, a fast and comprehensive research tool for multiple stages of signaling pathways has been developed. The open character of the cell-free system enables direct adjustments, modifications and analysis of each step of protein-protein interactions. In combination with high throughput screening approaches, this platform represents a sophisticated module for future drug development. A proof of concept for designing metabolic pathways in cell-free systems is already shown for *E. coli* based cell-free systems addressing the field of industrial biotechnology and biofuel production [56–58]. Future developments will focus on eukaryotic cell-free systems to implement this technology into drug development approaches.

In addition to WNT proteins, another therapeutically relevant protein, hTERT, was produced in three different eukaryotic cell-free systems. Ribonucleoprotein hTERT consists of a telomerase protein component (TERT) and a telomerase RNA sequence (TR) harboring only two conserved structural elements, the pseudoknot-template-region and the hairpin structure [27]. To obtain functionally active hTERT, both components need to be available in the protein expression system. The open nature of cell-free protein synthesis enables the direct supplementation of TR, which simplifies the production of the active enzyme. Our results demonstrate that even the cell-free synthesis of complex enzymes is feasible. In this study, two out of three analyzed eukaryotic cell-free systems enabled the production of active telomerase. Noticeable is the high background signal of CHO cell lysate control samples, which might be due to the presence of endogenous TERT in this particular system. Former studies showed that oligodeoxynucleotides containing human telomeric sequences hybridize with telomeres of a huge number of tested vertebrates [59]. Moreover, a cross-hybridization to insect and plant telomeres applying stringent conditions was not observed. Both findings indicate that human telomerase reverse transcriptase can rarely recognize a TR-component that is derived from non-vertebrate organisms. In this context, detectable activity of synthesized telomerase in CHO cell lysates might be possible without applying any additional TR since endogenous TR might be present in these lysates. Diverse production systems for hTERT have been available until now. The synthesis of human telomerase in a eukaryotic cell-free system based on rabbit reticulocyte cells was previously demonstrated by Bachand et al. [60]. The reconstitution of hTERT and a previously transcribed TR resulted in an active telomerase complex. A second alternative, *S. cerevisiae* cells with co-expressed hTERT and hTR, was utilized as a basis to perform functional studies [48]. Both systems provide a powerful tool to study hTERT and hTR interactions in vitro. Nevertheless, for therapeutical approaches which require high yields of active enzyme, both systems have their limitations since relatively low protein yields in the range of several µg/mL were obtained. The use of eukaryotic cell-free systems introduces novel opportunities for the future of hTERT and for the development of potential clinical applications in this area. hTERT is involved in various cancer types [61], therefore further characterization of the hTERT variants and targeted development of drugs may significantly improve cancer treatments.

In this study, the advantages of eukaryotic cell-free systems in the production of druggable protein targets were depicted. Production of target proteins in different eukaryotic cell-free systems was demonstrated showing the diversity of this platform technology. The selection of the type of eukaryotic cell-free system is thereby dependent on the target protein class and the requirement for the desired application. Wheat germ cell-free systems are typically limited in the performance of certain posttranslational modifications [18] and therefore not suitable for complex, modified mammalian proteins, but enable comparable high protein yields. *Sf*21 and CHO extract based cell-free systems allow for the synthesis and modification of complex target proteins and membrane-embedded proteins but are comparatively limited in their productivity. The improvements of eukaryotic cell-free systems in combination with the potential for screening applications and fast and easy protein production opens up new perspectives in future drug development. Initial approaches for the development of miniaturized platforms for high throughput cell-free synthesis are already reported [62,63]. First studies showed screening systems based on *E.coli* and wheat germ cell-free systems [64,65]. Expanding screening technologies for a broad variety of eukaryotic cell-free system will promote future drug development.

Supplementary Materials: The following are available online at http://www.mdpi.com/2409-9279/2/2/30/s1, Figure S1: Cell-free expression of WNT-Signalling Pathway receptor ligands WNT3a, WNT5a, WNT5b, WNT6 based on insect cell lysate and human cell lysat.

Author Contributions: Conceptualization, L.T., A.Z., D.W. and S.K.; methodology, L.T., A.Z., and D.W.; investigation, L.T., A.Z.; writing—review and editing L.T., A.Z. and S.K.; supervision, S.K.; funding acquisition S.K.

Funding: This work is supported by the European Regional Development Fund (EFRE) and the German Ministry of Education and Research (BMBF, No. 031B0078A).

Acknowledgments: The authors would like to thank Dana Wenzel for the preparation of cell lysates (Fraunhofer IZI, Potsdam-Golm, Germany).

References

1. Moorkens, E.; Meuwissen, N.; Huys, I.; Declerck, P.; Vulto, A.G.; Simoens, S. The Market of Biopharmaceutical Medicines: A Snapshot of a Diverse Industrial Landscape. *Front. Pharmacol.* **2017**, *8*, 314. [CrossRef]
2. Walsh, G. Biopharmaceutical benchmarks 2018. *Nat. Biotechnol.* **2018**, *36*, 1136–1145. [CrossRef]
3. Hughes, J.P.; Rees, S.; Kalindjian, S.B.; Philpott, K.L. Principles of early drug discovery. *Br. J. Pharmacol.* **2011**, *162*, 1239–1249. [CrossRef]
4. Zhang, Z.; Tang, W. Drug metabolism in drug discovery and development. *Acta Pharm. Sin. B* **2018**, *8*, 721–732. [CrossRef] [PubMed]
5. Bull, S.C.; Doig, A.J. Properties of Protein Drug Target Classes. *PLoS ONE* **2015**, *10*, e0117955. [CrossRef] [PubMed]
6. Bergauer, T.; Ruppert, T.; Essioux, L.; Spleiss, O. Drug Target Identification and Validation: Global Pharmaceutical Industry Experts on Challenges, Best Strategies, Innovative Precompetitive Collaboration Concepts, and Future Areas of Industry Precompetitive Research and Development. *Ther. Innov. Regul. Sci.* **2016**, *50*, 769–776. [CrossRef] [PubMed]
7. Jones, L.H. An industry perspective on drug target validation. *Expert Opin. Drug Discov.* **2016**, *11*, 623–625. [CrossRef] [PubMed]
8. Sugiki, T.; Fujiwara, T.; Kojima, C. Latest approaches for efficient protein production in drug discovery. *Expert Opin. Drug Discov.* **2014**, *9*, 1189–1204. [CrossRef]
9. Cuozzo, J.W.; Soutter, H.H. Overview of Recent Progress in Protein-Expression Technologies for Small-Molecule Screening. *J. Biomol. Screen.* **2014**, *19*, 1000–1013. [CrossRef]
10. Lu, Y. Cell-free synthetic biology: Engineering in an open world. *Synth. Syst. Biotechnol.* **2017**, *2*, 23–27. [CrossRef]
11. Nirenberg, M.W.; Matthaei, J.H. The dependence of cell-free protein synthesis in E. coli upon naturally occurring or synthetic polyribonucleotides. *Proc. Natl. Acad. Sci. USA* **1961**, *47*, 1588–1602. [CrossRef] [PubMed]
12. Chong, S. Overview of cell-free protein synthesis: Historic landmarks, commercial systems, and expanding applications. *Curr. Protoc. Mol. Biol.* **2014**, *108*, 16–30. [CrossRef] [PubMed]
13. He, M.; Taussig, M.J. Production of protein arrays by cell-free systems. *Methods Mol. Biol. (Clifton N.J.)* **2008**, *484*, 207–215. [CrossRef]
14. Díez, P.; González-González, M.; Lourido, L.; Dégano, R.M.; Ibarrola, N.; Casado-Vela, J.; LaBaer, J.; Fuentes, M. NAPPA as a Real New Method for Protein Microarray Generation. *Microarrays* **2015**, *4*, 214–227. [CrossRef] [PubMed]
15. Yu, X.; Song, L.; Petritis, B.; Bian, X.; Wang, H.; Viloria, J.; Park, J.; Bui, H.; Li, H.; Wang, J.; et al. Multiplexed Nucleic Acid Programmable Protein Arrays. *Theranostics* **2017**, *7*, 4057–4070. [CrossRef] [PubMed]
16. Littlefield, J.W.; Keller, E.B.; Gross, J.; Zamecnik, P.C. Studies on cytoplasmic ribonucleoprotein particles from the liver of the rat. *J. Biol. Chem.* **1955**, *217*, 111–123.
17. Gonano, F.; Baglioni, C. Initiation of hemoglobin synthesis with rabbit and E. coli tRNA in a reticulocyte cell-free system. *Eur. J. Biochem.* **1969**, *11*, 7–11. [CrossRef]
18. Zemella, A.; Thoring, L.; Hoffmeister, C.; Kubick, S. Cell-Free Protein Synthesis: Pros and Cons of Prokaryotic and Eukaryotic Systems. *ChemBioChem* **2015**, *16*, 2420–2431. [CrossRef]
19. Wu, C.; Sachs, M.S. Preparation of a Saccharomyces cerevisiae cell-free extract for in vitro translation. *Methods Enzymol.* **2014**, *539*, 17–28. [CrossRef]
20. Gan, R.; Jewett, M.C. A combined cell-free transcription-translation system from Saccharomyces cerevisiae for rapid and robust protein synthe. *Biotechnol. J.* **2014**, *9*, 641–651. [CrossRef]
21. Kubick, S.; Schacherl, J.; Fleischer-Notter, H.; Royall, E.; Roberts, L.O.; Stiege, W. In Vitro Translation in an Insect-Based Cell-Free System. In *Cell-Free Protein Expression*; Swartz, J.R., Ed.; Springer: Berlin/Heidelberg, Germany, 2003; pp. 209–217.
22. Thoring, L.; Dondapati, S.K.; Stech, M.; Wüstenhagen, D.A.; Kubick, S. High-yield production of "difficult-to-express" proteins in a continuous exchange cell-free system based on CHO cell lysates. *Sci. Rep.* **2017**, *7*, 11710. [CrossRef] [PubMed]
23. Buntru, M.; Vogel, S.; Spiegel, H.; Schillberg, S. Tobacco BY-2 cell-free lysate: An alternative and highly-productive plant-based in vitro translation system. *BMC Biotechnol.* **2014**, *14*, 37. [CrossRef] [PubMed]

24. Yadavalli, R.; Sam-Yellowe, T. HeLa Based Cell Free Expression Systems for Expression of Plasmodium Rhoptry Proteins. *J. Vis. Exp. JoVE* **2015**, e52772. [CrossRef] [PubMed]

25. Herr, P.; Hausmann, G.; Basler, K. WNT secretion and signalling in human disease. *Trends Mol. Med.* **2012**, *18*, 483–493. [CrossRef] [PubMed]

26. Orth, J.H.C.; Schorch, B.; Boundy, S.; Ffrench-Constant, R.; Kubick, S.; Aktories, K. Cell-free synthesis and characterization of a novel cytotoxic pierisin-like protein from the cabbage butterfly Pieris rapae. *Toxicon* **2011**, *57*, 199–207. [CrossRef]

27. Sandin, S.; Rhodes, D. Telomerase structure. *Curr. Opin. Struct. Biol.* **2014**, *25*, 104–110. [CrossRef]

28. Stech, M.; Quast, R.B.; Sachse, R.; Schulze, C.; Wüstenhagen, D.A.; Kubick, S. A continuous-exchange cell-free protein synthesis system based on extracts from cultured insect cells. *PLoS ONE* **2014**, *9*, e96635. [CrossRef]

29. Kuhl, S.J.; Kuhl, M. On the role of Wnt/beta-catenin signaling in stem cells. *Biochim. Biophys. Acta* **2013**, *1830*, 2297–2306. [CrossRef]

30. Teo, J.-L.; Kahn, M. The Wnt signaling pathway in cellular proliferation and differentiation: A tale of two coactivators. *Adv. Drug Deliv. Rev.* **2010**, *62*, 1149–1155. [CrossRef]

31. Camilli, T.C.; Weeraratna, A.T. Striking the target in Wnt-y conditions: Intervening in Wnt signaling during cancer progression. *Biochem. Pharmacol.* **2010**, *80*, 702–711. [CrossRef]

32. Inestrosa, N.C.; Toledo, E.M. The role of Wnt signaling in neuronal dysfunction in Alzheimer's Disease. *Mol. Neurodegener.* **2008**, *3*, 9. [CrossRef]

33. MacDonald, B.T.; Tamai, K.; He, X. Wnt/beta-catenin signaling: Components, mechanisms, and diseases. *Dev. Cell* **2009**, *17*, 9–26. [CrossRef]

34. Wilson, C. Diabetes: Human beta-cell proliferation by promoting Wnt signalling. *Nat. Rev. Endocrinol.* **2013**, *9*, 502. [CrossRef]

35. Cajanek, L.; Adlerz, L.; Bryja, V.; Arenas, E. WNT unrelated activities in commercially available preparations of recombinant WNT3a. *J. Cell. Biochem.* **2010**, *111*, 1077–1079. [CrossRef]

36. Green, J.L.; Bauer, M.; Yum, K.W.; Li, Y.-C.; Cox, M.L.; Willert, K.; Wahl, G.M. Use of a molecular genetic platform technology to produce human Wnt proteins reveals distinct local and distal signaling abilities. *PLoS ONE* **2013**, *8*, e58395. [CrossRef]

37. Kubick, S.; Gerrits, M.; Merk, H.; Stiege, W.; Erdmann, V.A. Chapter 2 In Vitro Synthesis of Posttranslationally Modified Membrane Proteins. *Curr. Top. Membr.* **2009**, *63*, 25–49.

38. Perez, J.G.; Stark, J.C.; Jewett, M.C. Cell-Free Synthetic Biology: Engineering Beyond the Cell. *Cold Spring Harb. Perspect. Biol.* **2016**, *8*, a023853. [CrossRef]

39. Caschera, F.; Noireaux, V. Preparation of amino acid mixtures for cell-free expression systems. *BioTechniques* **2015**, *58*, 40–43. [CrossRef]

40. Kim, D.-M.; Kim, Y.-E.; Choi, C.-Y. Enhancement of protein synthesis with sodium ion in a cell-free system from Escherichia coli. *J. Ferment. Bioeng.* **1996**, *82*, 398–400. [CrossRef]

41. Brigotti, M.; Petronini, P.G.; Carnicelli, D.; Alfieri, R.R.; Bonelli, M.A.; Borghetti, A.F.; Wheeler, K.P. Effects of osmolarity, ions and compatible osmolytes on cell-free protein synthesis. *Biochem. J.* **2003**, *369*, 369–374. [CrossRef]

42. Mikami, S.; Masutani, M.; Sonenberg, N.; Yokoyama, S.; Imataka, H. An efficient mammalian cell-free translation system supplemented with translation factors. *Protein Expr. Purif.* **2006**, *46*, 348–357. [CrossRef]

43. Brodiazhenko, T.; Johansson, M.J.O.; Takada, H.; Nissan, T.; Hauryliuk, V.; Murina, V. Elimination of Ribosome Inactivating Factors Improves the Efficiency of Bacillus subtilis and Saccharomyces cerevisiae Cell-Free Translation Systems. *Front. Microbiol.* **2018**, *9*, 3041. [CrossRef]

44. Axten, J.M.; Medina, J.R.; Feng, Y.; Shu, A.; Romeril, S.P.; Grant, S.W.; Li, W.H.H.; Heerding, D.A.; Minthorn, E.; Mencken, T.; et al. Discovery of 7-methyl-5-(1-{3-(trifluoromethyl)phenylacetyl)-2,3-dihydro-1H-indol-5-yl)-7H-p yrrolo2,3-dpyrimidin-4-amine (GSK2606414), a potent and selective first-in-class inhibitor of protein kinase R (PKR)-like endoplasmic reticulum kinase (PERK). *J. Med. Chem.* **2012**, *55*, 7193–7207. [CrossRef]

45. Szegezdi, E.; Logue, S.E.; Gorman, A.M.; Samali, A. Mediators of endoplasmic reticulum stress-induced apoptosis. *EMBO Rep.* **2006**, *7*, 880–885. [CrossRef]

46. Hong, S.H.; Ntai, I.; Haimovich, A.D.; Kelleher, N.L.; Isaacs, F.J.; Jewett, M.C. Cell-free protein synthesis from a release factor 1 deficient Escherichia coli activates efficient and multiple site-specific nonstandard amino acid incorporation. *ACS Synth. Biol.* **2014**, *3*, 398–409. [CrossRef]

47. Mihara, E.; Hirai, H.; Yamamoto, H.; Tamura-Kawakami, K.; Matano, M.; Kikuchi, A.; Sato, T.; Takagi, J. Active and water-soluble form of lipidated Wnt protein is maintained by a serum glycoprotein afamin/α-albumin. *eLife* **2016**, *5*, e11621. [CrossRef]

48. Bachand, F.; Autexier, C. Functional reconstitution of human telomerase expressed in Saccharomyces cerevisiae. *J. Biol. Chem.* **1999**, *274*, 38027–38031. [CrossRef]

49. Masutomi, K.; Kaneko, S.; Hayashi, N.; Yamashita, T.; Shirota, Y.; Kobayashi, K.; Murakami, S. Telomerase activity reconstituted in vitro with purified human telomerase reverse transcriptase and human telomerase RNA component. *J. Biol. Chem.* **2000**, *275*, 22568–22573. [CrossRef]

50. Nusse, R.; Clevers, H. Wnt/beta-Catenin Signaling, Disease, and Emerging Therapeutic Modalities. *Cell* **2017**, *169*, 985–999. [CrossRef]

51. Tai, D.; Wells, K.; Arcaroli, J.; Vanderbilt, C.; Aisner, D.L.; Messersmith, W.A.; Lieu, C.H. Targeting the WNT Signaling Pathway in Cancer Therapeutics. *Oncologist* **2015**, *20*, 1189–1198. [CrossRef]

52. Stech, M.; Merk, H.; Schenk, J.A.; Stocklein, W.F.M.; Wustenhagen, D.A.; Micheel, B.; Duschl, C.; Bier, F.F.; Kubick, S. Production of functional antibody fragments in a vesicle-based eukaryotic cell-free translation system. *J. Biotechnol.* **2012**, *164*, 220–231. [CrossRef]

53. Le Fourn, V.; Girod, P.-A.; Buceta, M.; Regamey, A.; Mermod, N. CHO cell engineering to prevent polypeptide aggregation and improve therapeutic protein secretion. *Metab. Eng.* **2014**, *21*, 91–102. [CrossRef]

54. Sachse, R.; Wüstenhagen, D.; Šamalíková, M.; Gerrits, M.; Bier, F.F.; Kubick, S. Synthesis of membrane proteins in eukaryotic cell-free systems. *Eng. Life Sci.* **2013**, *13*, 39–48. [CrossRef]

55. Zhao, L.; Rooker, S.M.; Morrell, N.; Leucht, P.; Simanovskii, D.; Helms, J.A. Controlling the in vivo activity of Wnt liposomes. *Methods Enzymol.* **2009**, *465*, 331–347. [CrossRef]

56. Karim, A.S.; Jewett, M.C. Chapter Two—Cell-Free Synthetic Biology for Pathway Prototyping. *Methods Enzymol.* **2018**, *608*, 31–57.

57. Korman, T.P.; Opgenorth, P.H.; Bowie, J.U. A synthetic biochemistry platform for cell free production of monoterpenes from glucose. *Nat. Commun.* **2017**, *8*, 15526. [CrossRef]

58. Takasuka, T.E.; Walker, J.A.; Bergeman, L.F.; Vander Meulen, K.A.; Makino, S.-I.; Elsen, N.L.; Fox, B.G. Cell-free translation of biofuel enzymes. *Methods Mol. Biol. (Clifton N.J.)* **2014**, *1118*, 71–95. [CrossRef]

59. Meyne, J.; Ratliff, R.L.; Moyzis, R.K. Conservation of the human telomere sequence (TTAGGG)n among vertebrates. *Proc. Natl. Acad. Sci. USA* **1989**, *86*, 7049–7053. [CrossRef]

60. Bachand, F.; Kukolj, G.; Autexier, C. Expression of hTERT and hTR in cis reconstitutes and active human telomerase ribonucleoprotein. *RNA (New York N.Y.)* **2000**, *6*, 778–784. [CrossRef]

61. Leão, R.; Apolónio, J.D.; Lee, D.; Figueiredo, A.; Tabori, U.; Castelo-Branco, P. Mechanisms of human telomerase reverse transcriptase (hTERT) regulation: Clinical impacts in cancer. *J. Biomed. Sci.* **2018**, *25*, 22. [CrossRef]

62. Georgi, V.; Georgi, L.; Blechert, M.; Bergmeister, M.; Zwanzig, M.; Wustenhagen, D.A.; Bier, F.F.; Jung, E.; Kubick, S. On-chip automation of cell-free protein synthesis: New opportunities due to a novel reaction mode. *Lab Chip* **2016**, *16*, 269–281. [CrossRef]

63. Khnouf, R.; Chapman, B.D.; Hugh Fan, Z. Fabrication optimization of a miniaturized array device for cell-free protein synthesis. *Electrophoresis* **2011**, *32*, 3101–3107. [CrossRef]

64. Zhu, B.; Mizoguchi, T.; Kojima, T.; Nakano, H. Ultra-High-Throughput Screening of an In Vitro-Synthesized Horseradish Peroxidase Displayed on Microbeads Using Cell Sorter. *PLoS ONE* **2015**, *10*, e0127479. [CrossRef]

65. Endo, Y.; Sawasaki, T. High-throughput, genome-scale protein production method based on the wheat germ cell-free expression system. *Biotechnol. Adv.* **2003**, *21*, 695–713. [CrossRef]

methods
and
protocols

MDPI

Review

Cell-Free Metabolic Engineering: Recent Developments and Future Prospects

Hye Jin Lim and Dong-Myung Kim *

Department of Chemical Engineering and Applied Chemistry, Chungnam National University, Daejeon 34134, Korea; yhjysy@naver.com
* Corresponding: dmkim@cnu.ac.kr; Tel.: +82-42-823-7692

Received: 26 March 2019; Accepted: 24 April 2019; Published: 30 April 2019

Abstract: Due to the ongoing crises of fossil fuel depletion, climate change, and environmental pollution, microbial processes are increasingly considered as a potential alternative for cleaner and more efficient production of the diverse chemicals required for modern civilization. However, many issues, including low efficiency of raw material conversion and unintended release of genetically modified microorganisms into the environment, have limited the use of bioprocesses that rely on recombinant microorganisms. Cell-free metabolic engineering is emerging as a new approach that overcomes the limitations of existing cell-based systems. Instead of relying on metabolic processes carried out by living cells, cell-free metabolic engineering harnesses the metabolic activities of cell lysates in vitro. Such approaches offer several potential benefits, including operational simplicity, high conversion yield and productivity, and prevention of environmental release of microorganisms. In this article, we review the recent progress in this field and discuss the prospects of this technique as a next-generation bioconversion platform for the chemical industry.

Keywords: cell-free metabolic engineering; cell-free protein synthesis; bioconversion platform

---

1. Introduction

Owing to recent advances in genetic and genomic engineering techniques, microbial cells are increasingly being used as self-replicating microreactors that can produce diverse materials from exogenously introduced genes [1,2]. However, the use of living cells often prevents us from harnessing their full synthetic power. Living systems operate only within narrow condition ranges, including temperature, salt concentration and solvent properties. Toxicity or metabolic burden also limit high-volume production of recombinant products. In addition, the interconnectedness of cellular metabolic pathways often reduces substrate flux into synthetic pathways, thus lowering product yield and conversion efficiencies. Most of these problems stem from the requirement of living cells to maintain balanced homeostasis [3]. Liebig's law of the minimum teaches us that deterioration of any essential cellular component can result in failure of the entire system, thus preventing the operation of the desired pathways.

In theory, many of these problems can be avoided by using the individual biological components specifically required to produce the target products. In fact, the use of purified biosynthetic machinery in cell-free systems has a long history that spans several decades. A prominent example is the use of purified recombinant DNA polymerase. Purified DNA polymerase can be used for many more tasks than it performs in living cells. In addition to its common use for rapid DNA amplification in thermal cyclers, the DNA synthesis activity of DNA polymerases has been widely used for many applications in combination with various reagents and conditions, including diagnostic techniques and genetic mutagenesis [4].

Cell-free use of biosynthetic machinery has also been expanded to protein production, which is more complicated and requires many enzymes and translational factors. These components were purified or extracted from cells and successfully reconstituted to produce recombinant proteins directed by genetic programming contained in the reaction mixtures [5–9]. Cell-free metabolic engineering is the latest addition to these recent efforts to harness cellular functions outside of cells and it involves the use of purified or crude enzymes to produce chemical compounds [10,11]. Liberated from the requirement of maintaining cellular viability and growth, cell-free metabolic engineering provides far greater design flexibility and wider operational conditions for synthetic metabolic pathways. Cell-free metabolic engineering systems also offer important benefits that cannot be attained using living cells, including quantitative and precise assessment of performance by direct sampling, rapid cycles of design-build-test iterations and the capability to use non-natural or non-biological components. While the concept of cell-free metabolism was introduced as early as 100 years ago with the demonstration of ethanol production in crude yeast lysate [12], the use of enzymes has long been relegated to an auxiliary role in the production of structurally complex intermediates via organic synthesis approaches. However, growing demand for cleaner and more efficient chemical processes along with notable advances in genetic engineering and enzyme technology have led to recognition of cell-free synthetic approaches as a promising method for synthesizing the diverse range of chemical compounds used in industrial implications.

This review summarizes recent efforts to harness the principle of cell-free synthesis to reproduce intracellular reaction pathways outside of a model system and to reach yields and productivity that are not achievable with current cell-based methods. In particular, our discussion focuses on two closely related topics: synthesis of enzymes that catalyze chemical conversion pathways with industrial implications and production of important chemicals via cell-free use of the necessary enzymes. We also discuss the potential to integrate cell-free enzyme synthesis and metabolic engineering to build DNA-programmed, cell-free metabolic engineering systems.

2. Cell-Free Protein Synthesis Systems

2.1. Development of Highly Productive Cell-Free Protein Synthesis Systems

Similar to the case of PCR, the operational convenience and productivity of cell-free protein synthesis approaches have evolved over the last decades. For example, extensive studies on the factors limiting conventional cell-free protein synthesis systems have revealed that a steady and continuous ATP supply is one of the most important requirements for efficient protein production [13–15]. Each step of the molecular process of protein synthesis (aminoacylation, transcription, and translation) consumes large amounts of ATP. This consumption leads to a rapid decrease in the ATP level in the reaction mixture and limits the productivity of conventional cell-free protein synthesis systems [16,17]. Use of high concentrations of ATP or other energy sources cannot easily address this problem because of the resulting accumulation of inorganic phosphate, which inhibits protein synthesis by chelating magnesium ions, an essential cofactor required for this process [18,19]. Early attempts to address this dilemma involved continuous ATP supplementation and removal of inorganic phosphate via forced pumping [20,21] or diffusional exchange (Figure 1) [17,22,23].

While these approaches markedly improved the duration of the reaction and thus the final target protein yield, they had the drawback of requiring complex devices and excessive amounts of reagents [24]. Therefore, different strategies have been developed to improve the ATP supply during batch cell-free protein synthesis reactions, while avoiding inorganic phosphate accumulation (Figure 2). In 1999, Kim and Swartz developed a method for sustained ATP supplementation in a batch cell-free protein synthesis system without inorganic phosphate accumulation. Instead of phosphate-containing energy sources, they used pyruvate as a phosphate-free energy source to regenerate ATP [16]. Because pyruvate is the final product of the glycolytic pathway, their results inspired the use of glucose and glycolytic intermediates as energy sources for regenerating the ATP required for protein synthesis [25,26].

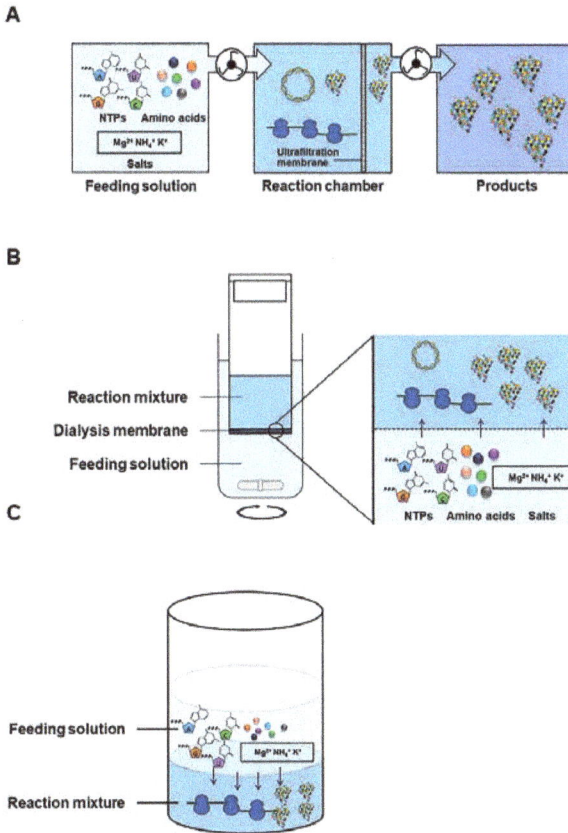

Figure 1. Reaction configurations of cell-free protein synthesis for continuous supply of substrates. (**A**) A continuous flow cell-free translation system. The feeding solution containing the substrates for protein synthesis continuously through the reaction mixture retained by an ultrafiltration membrane. (**B**) A continuous exchange cell-free protein synthesis system. The supply of substrates and removal of by-products are achieved by diffusional exchange through a dialysis membrane. (**C**) A bilayer cell-free protein synthesis system. Feeding solution is overlaid on top of the reaction mixture for cell-free protein synthesis and diffusional exchange of substrates and by-products take place at the interface of the two phases.

In subsequent studies, the ATP regeneration efficiency was further improved by using polymeric glucose as a high-density energy source [27,28]. Most recently, Caschera and Noireaux demonstrated that polyphosphate can be used as a highly efficient and cost-effective energy source for high-yield protein production in a cell-free protein synthesis system [29]. As a result of these efforts, production of milligram quantities of recombinant proteins in batch reactions is now routinely reported.

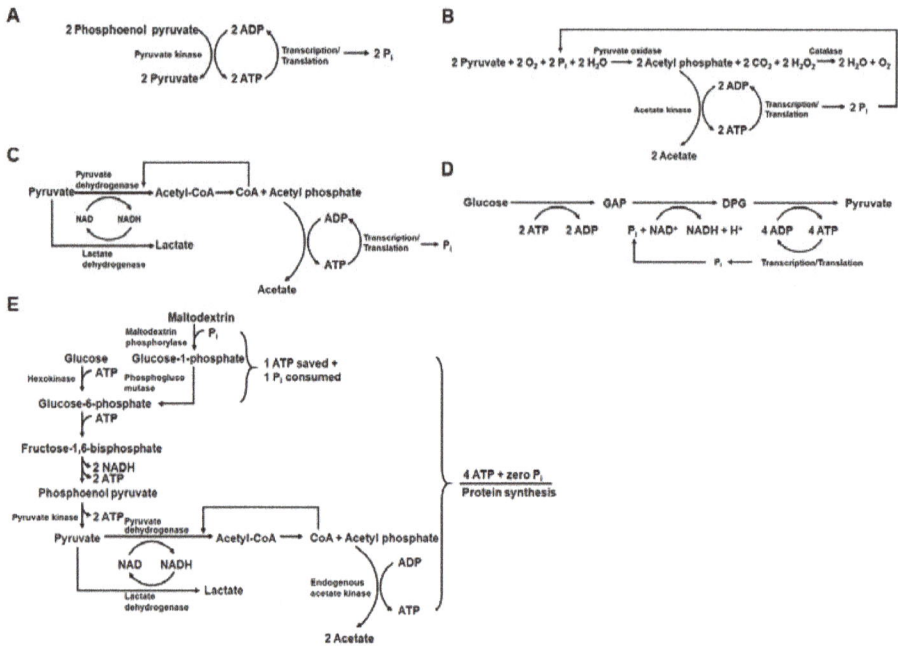

Figure 2. Biochemical strategies for enhanced supply of ATP without the accumulation of inorganic phosphate. (**A**) Conventional methods for ATP regeneration during cell-free protein synthesis simply rely on the substrate-level phosphorylation of ADP using the energy sources with high-energy phosphate bonds (i.e., phosphoenolpyruvate, creatine phosphate, and acetyl phosphate). In this scheme, inorganic phosphate accumulates in the reaction mixture in amounts proportional to those of the energy sources. (**B**) Kim and Swartz demonstrated that pyruvate can be used as a phosphate-free energy source, in combination with exogenously added pyruvate oxidase [16]. In their scheme, pyruvate and recycled inorganic phosphate produce acetyl phosphate, which is subsequently used to regenerate ATP. (**C**) Generation of acetyl phosphate from pyruvate can also be achieved by using the endogenous enzymes in the cell extract. (**D**) Use of glucose to regenerate ATP via glycolytic pathway in the cell extract. (**E**) Use of maltodextrin as a secondary energy source. ADP: adenosine diphosphate; ATP: adenosine triphosphate; Pi: inorganic phosphate; NAD: nicotinamide adenine dinucleotide; NADH: reduced nicotinamide adenine dinucleotide; GAP: glyceraldehyde-3-phosphate; DPG: 1,3-diphosphoglycerate.

2.2. Direct Programming of Cell-Free Protein Synthesis with Linear DNA Templates

Another important issue in cell-free protein synthesis is the method used to prepare the template DNA. When cell-free protein synthesis is directed by plasmid-borne genes (like in cell-based expression systems), it is still necessary to grow cells for cloning and template DNA amplification. This requirement off-sets the benefits of cell-free protein synthesis; however, the need for cell growth can be avoided by using PCR to prepare the template DNA. In general, the efficiency of cell-free synthesis directed by PCR-amplified DNA is substantially lower than that in reactions containing plasmid-borne templates. This effect is mainly due to rapid degradation of the linear templates by exonucleases present in the cell-free extract [30]. Numerous attempts have been made to address the issue of template DNA stability during cell-free protein synthesis (Figure 3). Sitaraman et al. developed a method to stabilize PCR-amplified linear DNA using the lambda phage Gam protein, which inhibits the RecBCD exonuclease [31]. Marshall et al. introduced Chi-sites into the template DNA to block RecBCD without the need for purified Gam protein [32]. Seki et al. improved the efficiency of cell-free protein synthesis from linear DNA templates via affinity-removal of polynucleotide phosphorylase (PHPase) and RecD from the cell extract [33]. Wu et al. designed stable linear templates cyclized between single-stranded 5′-phosphorylated overhangs by the endogenous ligase activity of *Escherichia coli* S30 extracts [34]. Ahn et al. took the alternative approach of stabilizing the mRNA transcribed from the linear DNA, rather than stabilizing the DNA itself. They found that the mRNA lifespan is remarkably extended when its 3′-end forms a stem-loop structure and the cell-free protein synthesis is conducted in an extract lacking RNase E activity. This enhanced mRNA stability, in turn, led to highly efficient protein expression at yields comparable to those obtained from reactions using plasmid templates [35]. By eliminating the time-consuming steps required for template preparation, these methods enable direct programming of cell-free protein synthesis systems for the instant production of the desired proteins.

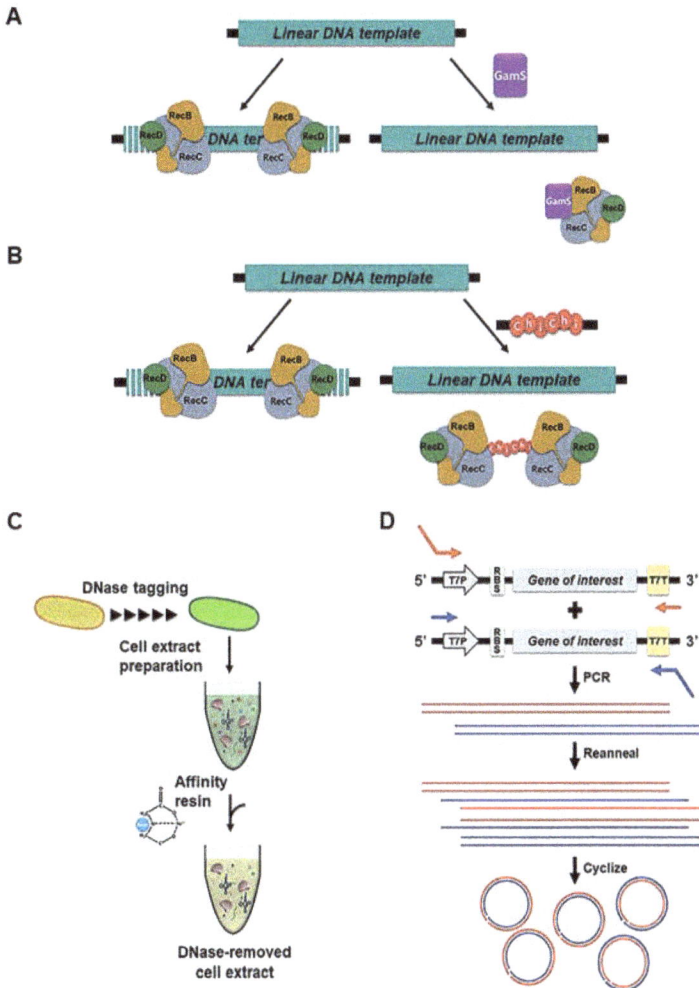

Figure 3. Strategies for improving the stability of linear DNA templates in cell-free protein synthesis systems. (**A**) Stabilization of linear template by use of GamS protein, an inhibitor of RecBCD complex. (**B**) Sequestration of RecBCD complex by using a short DNA containing repeated χ-site. (**C**) Affinity-removal of DNases from the cell extract. (**D**) Cyclization of linear templates by endogenous DNA ligase in the cell extract.

2.3. Cell-Free Enzyme Synthesis

As described above, cell-free protein synthesis techniques have rapidly evolved to produce large amounts of recombinant proteins directly from in vitro-produced template DNA. These advances have been successfully combined with the versatile nature of cell-free protein synthesis to produce enzymes that are otherwise difficult to express in functional forms [36] (Figure 4).

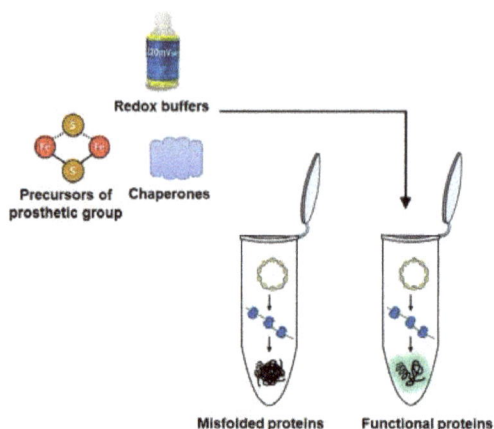

Figure 4. Enhanced production of functional enzymes by direct additions of folding effectors. The open nature of cell-free protein synthesis allows direct additions of various chemicals and biomolecules that assist proper folding of desired enzymes.

For example, functional *Candida antarctica* lipase B can be produced by simply adjusting the redox potential of the cell-free protein synthesis reaction mixture to facilitate intramolecular disulfide bond formation [37,38]. Cell-free synthesis also allows facile introduction of unnatural amino acids into an enzyme structure, which is particularly useful for producing enzymes that can be immobilized in controlled orientations. For example, through site-specific introduction of an unnatural amino acid containing a chemical handle, Wu et al. could immobilize T4 lysozyme on solid beads. It was found that optimal orientation of immobilization allowed substantially enhanced activity and stability of the immobilized enzyme [39]. Swartz et al. demonstrated the versatility of cell-free protein synthesis systems for producing complex enzymes by expressing functional [FeFe] hydrogenase. They could produce and mature algal and bacterial hydrogenases using *E. coli* extracts containing the HydG, HydE, and HydF proteins. Pre-incubation of these proteins with sulfide and iron in the reaction mixture allowed proper assembly of the iron-sulfur cluster and apoenzyme during the subsequent hydrogenase synthesis [40]. In a similar approach, Li et al. successfully expressed functional multicopper oxidase, which has potential biotechnological applications. This enzyme commonly shows low expression levels in traditional recombinant hosts; however, by simply adding copper sulfate to the reaction, the cell-free protein synthesis system yielded over 1 mg mL^{-1} of soluble and functional multicopper oxidase [41]. Kwon et al. used P450 BM3 as a proof-of-concept model to show that the pathways for prosthetic group and apoenzyme synthesis could be combined in a one-pot reaction to produce functional monooxygenase [42]. Cell-free protein synthesis systems are also a promising platform for producing enzymes that are toxic to recombinant hosts. For example, Lim et al. reported successful phospholipase A1 production using a cell-free protein synthesis system derived from E. coli. Phospholipase A1 degrades phospholipids in the cell membrane and, thus, cannot be efficiently produced in the cytoplasm of live *E. coli* cells. By decoupling enzyme expression from cell physiology in a cell-free protein synthesis system, they achieved an over 1000-fold higher yield of functional phospholipase A1 [43].

3. Cell-Free Metabolic Engineering

3.1. Purified Protein-Based Cell-Free Metabolic Engineering

The conditions for bioconversion using living cells are restricted by the physiological limits required to maintain life. In most cases, for example, microbial processes can only be operated below 40 °C to prevent cellular damage [44]. Furthermore, metabolic pathways that involve intermediates or products that are toxic to the host cells cannot be easily used. The complexity of cellular metabolism is another barrier for efficient target compound production. Due to these features, most natural microbes are not efficient enough to support high-yield production of target chemicals sufficient to meet the demands of current petroleum-based markets [45]. The most straightforward solution to these limitations is to use purified enzymes to build cell-free metabolic pathways. Cell-free metabolic pathways based on purified enzymes enable simple interpretation of results and optimization of the participating enzymes. For example, Bujara et al. successfully demonstrated how cell-free metabolic pathways can be optimized by coupling them to real-time analysis methods [46]. Cell-free enzymatic approaches allow flexible design of novel pathways solely focused on target molecule production. In an effort to eliminate the ATP-driven reactions required for the conversion of glucose into pyruvate, Guterl et al. developed an artificial glycolytic pathway that requires only four enzymes. Their simplified pyruvate synthesis pathway was subsequently streamlined via addition of enzymatic pathways for ethanol and isobutanol synthesis [47]. Despite the advantage of being free from the constraints imposed by cells, a shortcoming of cell-free metabolic pathways is that they are disconnected from the cellular biochemical replenishing systems. Korman et al. introduced a modular design for an artificial 27-enzyme pathway for cell-free monoterpene production. Through smart design and arrangement of the modules of the enzymatic pathways that produce intermediates and regenerate co-factors, such as ATP and NADPH, they built a balanced cell-free monoterpene synthesis process with a conversion yield of greater than 95% and titers greater than 15 g L^{-1} [10].

These results clearly demonstrate the potential of cell-free metabolic engineering for the production of industrial chemicals via artificial pathways. In addition to chemical production, Martin et al. developed a synthetic pathway that produces 10 moles of dihydrogen via consumption of one mole of ATP during xylose breakdown [48]. Furthermore, cell-free enzymatic pathways have also been successfully used for bioelectricity production. Zhu et al. reported a synthetic cell-free pathway that produces nearly 24 electrons per glucose unit in an aerobic enzymatic fuel cell. This enzymatic fuel cell exhibited an energy-storage density one order of magnitude higher than that of lithium-ion batteries [49].

3.2. Cell Extract-Based Cell-Free Metabolic Engineering

Despite the attractive advantages of enzyme-based cell-free pathways, the requirement for laborious purification of individual enzymes limits their use, particularly for multistep reactions. In cases where the intermediates of the final products can be generated via cellular metabolism, a meet-in-the-middle strategy employing cell lysates might be a more realistic approach [50–53]. Such an approach would harness the activities of the cellular components after removal of the membrane barrier and complement the cellular pathway that converts a raw material into the necessary intermediates by introducing additional enzymes required to generate the final product (Figure 5). For example, Bujara et al. produced a series of unnatural monosaccharides from glucose by adding enzymes to the E. coli lysate that complete the final synthesis steps. The cell-free metabolism intrinsic to the E. coli lysate resulted in accumulation of dihydroxyacetone phosphate (DHAP), which was subsequently converted into unnatural monosaccharides via the actions of exogenously added enzymes [54]. In a similar approach, Kay and Jewett established a cell-free metabolic pathway for 2,3-butanediol production. In their study, the pyruvate synthesized by the cell-free glycolytic pathway was then successfully converted to 2,3-butanediol via acetolactate and acetoin through exogenous addition of the required enzymes (acetolactate synthase, acetolactate decarboxylase and butanediol dehydrogenase) to the reaction mixture [11]. The titer of the cell-free-synthesized 2,3-butanediol reached 80 g L^{-1}, which is close to the theoretical yield. Yi et al. proposed an interesting alternative approach for 2,3-butanediol synthesis based on a hybrid cell-free synthesis system. The cyanobacterial endogenous starch-breakdown pathway was combined with the E. coli glycolytic pathway by mixing lysates of the two species. As this mixed-lysate system accumulated pyruvate from starch, addition of acetolactate synthase, acetolactate decarboxylase and butanediol dehydrogenase led to successful 2,3-butanediol synthesis. These results demonstrate that the synthesis of new heterologous metabolic pathways could support biomolecule synthesis [55].

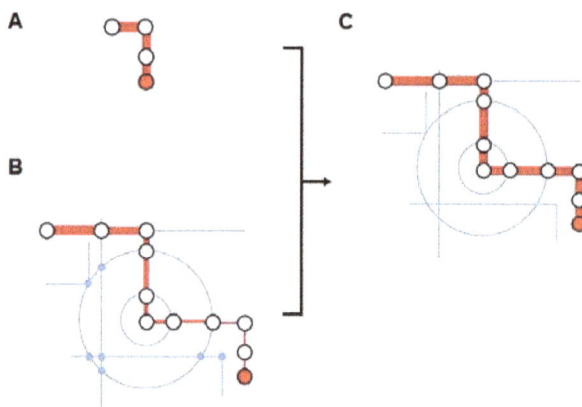

Figure 5. Cell-free metabolic engineering based on cell extract. (**A**) Enzyme-based cell-free metabolic engineering offers greater flexibility and controllability. However, the requirement for purified individual enzyme limits its use for multi-step bioconversion. (**B**) On the other hand, cell-based metabolic engineering suffers from the complexity of cellular metabolic pathways and the fluxes of intermediates can be reduced by their cellular consumptions (represented by gradual reduction of the thickness of red line). (**C**) Cell-free metabolic engineering can hitchhike the cellular metabolism, while minimizing genetic modification of host cells and loss of intermediates.

4. Conclusions

Cell-free metabolic engineering is expected to provide an alternative route for biological production of chemical compounds. As systems developed through cell-free metabolic engineering are independent of cell viability and growth and insulated from toxicity of the synthesized chemicals, they can offer increased flexibility and higher conversion efficiency. While most of the studies on this topic, including those discussed in this review, have been conducted using purified or extracted enzymes, an interesting approach would be to integrate cell-free metabolic engineering with PCR and cell-free protein synthesis to establish a directly programmable metabolic engineering platform [56]. For this concept to be developed into a practical method, additional techniques will be needed, including methods for expressing functional proteins and for regulating the expression levels of exogenous enzymes. Considering the marked progress in the development of such techniques [57–59], it is likely that genetically programmed and controlled cell-free metabolic engineering platforms will soon emerge.

Author Contributions: Conceptualization, D.-M.K.; Writing-Original Draft Preparation, H.J.L.; Writing-Review & Editing, D.-M.K. and H.J.L.; Visualization, D.-M.K. and H.J.L.; Supervision, D.-M.K.; Funding Acquisition, D.-M.K.

Funding: This research was funded by National Research Foundation of Korea (grant numbers 2015M3D3A1A01064878 and 2016M1A5A1027465.

Conflicts of Interest: Authors declare no conflict of interest.

References

1. McCarty, N.S.; Ledesma-Amaro, R. Synthetic biology tools to engineer microbial communities for biotechnology. *Trends Biotechnol.* **2019**, *37*, 181–197. [CrossRef] [PubMed]
2. Vervoort, Y.; Linares, A.G.; Roncoroni, M.; Liu, C.; Steensels, J.; Verstrepen, K.J. High-throughput system-wide engineering and screening for microbial biotechnology. *Curr. Opin. Biotechnol.* **2017**, *46*, 120–125. [CrossRef]
3. Dudley, Q.M.; Karim, A.S.; Jewett, M.C. Cell-free metabolic engineering: Biomanufacturing beyond the cell. *Biotechnol. J.* **2015**, *10*, 69–82. [CrossRef]
4. Morley, A.A. Digital PCR: A brief history. *Biomol. Detect. Quantif.* **2014**, *1*, 1–2. [CrossRef] [PubMed]
5. Shimizu, Y.; Inoue, A.; Tomari, Y.; Suzuki, T.; Yokogawa, T.; Nishikawa, K.; Ueda, T. Cell-free translation reconstituted with purified components. *Nat. Biotechnol.* **2001**, *19*, 751–755. [CrossRef]
6. Kuruma, Y.; Ueda, T. The PURE system for the cell-free synthesis of membrane proteins. *Nat. Protoc.* **2015**, *10*, 1328–1344. [CrossRef] [PubMed]
7. Li, J.; Gu, L.; Aach, J.A.; Church, G.M. Improved cell-free RNA and protein synthesis system. *PLoS ONE* **2014**, *9*, e106232. [CrossRef]
8. Carlson, E.D.; Gan, R.; Hodgman, C.E.; Jewett, M.C. Cell-free protein synthesis: Applications come of age. *Biotechnol. Adv.* **2012**, *30*, 1185–1194. [CrossRef]
9. Ahn, J.H.; Hwang, M.Y.; Lee, K.H.; Choi, C.Y.; Kim, D.M. Use of signal sequences as an in situ removable sequence element to stimulate protein synthesis in cell-free extracts. *Nucleic Acids Res.* **2007**, *35*, e21. [CrossRef]
10. Korman, T.P.; Opgenorth, P.H.; Bowie, J.U. A synthetic biochemistry platform for cell free production of monoterpenes from glucose. *Nat. Commun.* **2017**, *8*, 15526. [CrossRef] [PubMed]
11. Kay, J.E.; Jewett, M.C. Lysate of engineered *Escherichia coli* supports high-level conversion of glucose to 2,3-butanediol. *Metab. Eng.* **2015**, *32*, 133–142. [CrossRef]
12. Buchner, E. Alkoholische gärung ohne hefezellen. *Ber. Chem. Ges.* **1897**, *30*, 117–124. [CrossRef]
13. Dopp, B.J.L.; Tamiev, D.D.; Reuel, N.F. Cell-free supplement mixtures: Elucidating the history and biochemical utility of additives used to support in vitro protein synthesis in *E. coli* extract. *Biotechnol. Adv.* **2019**, *37*, 246–258. [CrossRef] [PubMed]
14. Jewett, M.C.; Swartz, J.R. Mimicking the *Escherichia coli* cytoplasmic environment activates long-lived and efficient cell-free protein synthesis. *Biotechnol. Bioeng.* **2004**, *88*, 19–26. [CrossRef] [PubMed]
15. Kim, H.C.; Kim, D.M. Methods for energizing cell-free protein synthesis. *J. Biosci. Bioeng.* **2009**, *1*, 1–4. [CrossRef] [PubMed]
16. Kim, D.M.; Swartz, J.R. Prolonging cell-free protein synthesis with a novel ATP regeneration system. *Biotechnol. Bioeng.* **1999**, *66*, 180–188. [CrossRef]
17. Kim, D.M.; Choi, C.Y. A semicontinuous prokaryotic coupled transcription/translation system using a dialysis membrane. *Biotechnol. Prog.* **1996**, *12*, 645–649. [CrossRef] [PubMed]
18. Kim, T.W.; Kim, D.M.; Choi, C.Y. Rapid production of milligram quantities of proteins in a batch cell-free protein synthesis system. *J. Biotechnol.* **2006**, 373–380. [CrossRef] [PubMed]
19. Whittaker, J.W. Cell-free protein synthesis: The state of the art. *Biotechnol. Lett.* **2013**, *35*, 143–152. [CrossRef]
20. Spirin, A.S.; Baranov, V.I.; Ryabova, L.A.; Ovodov, S.Y.; Alakhov, Y.B. A continuous cell-free translation system capable of producing polypeptides in high yield. *Science* **1988**, *242*, 1162–1164. [CrossRef]
21. Endo, Y.; Otsuzuki, S.; Ito, K.; Miura, K.-I. Production of an enzymatic active protein using a continuous flow cell-free translation system. *J. Biotechnol.* **1992**, *25*, 221–230. [CrossRef]
22. Chen, H.; Xu, Z.; Xu, N.; Cen, P. Efficient production of a soluble fusion protein containing human beta-defensin-2 in *E. coli* cell-free system. *J. Biotechnol.* **2005**, *115*, 307–315. [CrossRef] [PubMed]
23. Go, S.Y.; Lee, K.H.; Kim, D.M. Detergent-assisted enhancement of the translation rate during cell-free synthesis of peptides in an *Escherichia coli* extract. *Biotechnol. Bioprocess Eng.* **2019**, *23*, 679–685. [CrossRef]
24. Spirin, A.S. High-throughput cell-free systems for synthesis of functionally active proteins. *Trends Biotechnol.* **2004**, *22*, 538–545. [CrossRef]
25. Kim, D.M.; Swartz, J.R. Regeneration of adenosine triphosphate from glycolytic intermediates for cell-free protein synthesis. *Biotechnol. Bioeng.* **2001**, *74*, 309–316. [CrossRef]
26. Calhoun, K.A.; Swartz, J.R. Energizing cell-free protein synthesis with glucose metabolism. *Biotechnol. Bioeng.* **2005**, *90*, 606–613. [CrossRef]

27. Wang, Y.; Zhang, Y.-H.P. Cell-free protein synthesis energized by slowly-metabolized maltodextrin. *BMC Biotechnol.* **2009**, *9*, 58. [CrossRef]
28. Kim, H.C.; Kim, T.W.; Kim, D.M. Prolonged production of proteins in a cell-free protein synthesis system using polymeric carbohydrates as an energy source. *Process Biochem.* **2011**, *46*, 1366–1369. [CrossRef]
29. Caschera, F.; Noireaux, V. A cost-effective polyphosphate-based metabolism fuels an all *E. coli* cell-free expression system. *Metab. Eng.* **2015**, *27*, 29–37. [CrossRef]
30. Lesley, S.A.; Brow, M.A.; Burgess, R.R. Use of in vitro protein synthesis from polymerase chain reaction-generated templates to study interaction of *Escherichia coli* transcription factors with core RNA polymerase and for epitope mapping of monoclonal antibodies. *J. Biol. Chem.* **1991**, *266*, 2632–2638.
31. Sitaraman, K.; Esposito, D.; Klarmann, G.; Le Grice, S.F.; Hartley, J.L.; Chatterjee, D.K. A novel cell-free protein synthesis system. *J. Biotechnol.* **2004**, *110*, 257–263. [CrossRef]
32. Marshall, R.; Maxwell, C.S.; Collins, S.P.; Beisel, C.L.; Noireaux, V. Short DNA containing χ sites enhances DNA stability and gene expression in *E. coli* cell-free transcription-translation systems. *Biotechnol. Bioeng.* **2017**, *114*, 2137–2141. [CrossRef]
33. Seki, E.; Matsuda, N.; Kigawa, T. Multiple inhibitory factor removal from an *Escherichia coli* cell extract improves cell-free protein synthesis. *J. Biosci. Bioeng.* **2009**, *108*, 30–35. [CrossRef]
34. Wu, P.S.C.; Ozawa, K.; Lim, S.P.; Vasudevan, S.G.; Dixon, N.E.; Otting, G. Cell-free transcription/translation from PCR-amplified DNA for high-throughput NMR studies. *Angew Chem. Int. Ed.* **2007**, *46*, 3356–3358. [CrossRef]
35. Ahn, J.H.; Chu, H.-S.; Kim, T.W.; Oh, I.S.; Choi, C.Y.; Hahn, G.H.; Park, C.G.; Kim, D.M. Cell-free synthesis of recombinant proteins from PCR-amplified genes at a comparable productivity to that of plasmid-based reactions. *Biochem. Biophys. Res. Commun.* **2005**, *338*, 1346–1352. [CrossRef]
36. Hunt, J.P.; Yang, S.O.; Wilding, K.M.; Bundy, B.C. The growing impact of lyophilized cell-free expression systems. *Bioengineered* **2017**, *8*, 325–330. [CrossRef]
37. Park, C.G.; Kim, T.W.; Oh, I.S.; Song, J.K.; Kim, D.M. Expression of functional *Candida antarctica* Lipase B in a cell-free protein synthesis system derived from *Escherichia coli*. *Biotechnol. Prog.* **2009**, *25*, 589–593. [CrossRef]
38. Park, C.G.; Kwon, M.A.; Song, J.K.; Kim, D.M. Cell-free synthesis and multifold screening of *Candida antarctica* Lipase B variants after combinatorial mutagenesis of hot spots. *Biotechnol. Prog.* **2011**, *27*, 47–53. [CrossRef]
39. Wu, J.C.Y.; Hutchings, C.H.; Lindsay, M.J.; Werner, C.J.; Bundy, B.C. Enhanced enzyme stability through site-directed covalent immobilization. *J. Biotechnol.* **2015**, *193*, 83–90. [CrossRef]
40. Boyer, M.E.; Stapleton, J.A.; Kuchenreuther, J.M.; Wang, C.W.; Swartz, J.R. Cell-free synthesis and maturation of [FeFe] hydrogenases. *Biotechnol. Bioeng.* **2008**, *99*, 59–67. [CrossRef]
41. Li, J.; Lawton, T.J.; Kostecki, J.S.; Nisthal, A.; Fang, J.; Mayo, S.L.; Rosenzweig, A.C.; Jewett, M.C. Cell-free protein synthesis enables high yielding synthesis of an active multicopper oxidase. *Biotechnol. J.* **2016**, *11*, 212–218. [CrossRef]
42. Kwon, Y.C.; Oh, I.S.; Lee, N.; Lee, K.H.; Yoon, Y.J.; Lee, E.Y.; Kim, B.G.; Kim, D.M. Integrating cell-free biosynthesis of heme prosthetic group and apoenzyme for the synthesis of functional P450 monooxygenase. *Biotechnol. Bioeng.* **2013**, *110*, 1193–1200. [CrossRef]
43. Lim, H.J.; Park, Y.J.; Jang, Y.J.; Choi, J.E.; Oh, J.Y.; Park, J.H.; Song, J.K.; Kim, D.M. Cell-free synthesis of functional phospholipase A1 from *Serratia* sp. *Biotechnol. Biofuels* **2016**, *9*, 159. [CrossRef]
44. Price, P.B.; Sowers, T. Temperature dependence of metabolic rates for microbial growth, maintenance and survival. *Proc. Natl. Acad. Sci. USA* **2004**, *10*, 4631–4636. [CrossRef]
45. Chubukov, V.; Mukhopadhyay, A.; Petzold, C.J.; Keasling, J.D.; Martín, H.G. Synthetic and systems biology for microbial production of commodity chemicals. *NPJ Syst. Biol. Appl.* **2016**, *2*, 16009. [CrossRef]
46. Bujara, M.; Schümperli, M.; Pellaux, R.; Heinemann, M.; Panke, S. Optimization of a blueprint for in vitro glycolysis by metabolic real-time analysis. *Nat. Chem. Biol.* **2011**, *7*, 271–277. [CrossRef]
47. Guterl, J.K.; Garbe, D.; Carsten, J.; Steffler, F.; Sommer, B.; Reiße, S.; Philipp, A.; Haack, M.; Rühmann, B.; Koltermann, A.; et al. Cell-free metabolic engineering: Production of chemicals by minimized reaction cascades. *ChemSusChem* **2012**, *5*, 2165–2172. [CrossRef]
48. Martín del Campo, J.S.; Rollin, J.; Myung, S.; Chun, Y.; Chandrayan, S.; Patiño, R.; Adams, M.W.; Zhang, Y.-H.P. High-yield production of dihydrogen from xylose by using a synthetic enzyme cascade in a cell-free system. *Angew. Chem. Int. Ed. Engl.* **2013**, *52*, 4587–4590.

49. Zhu, Z.; Tam, T.K.; Sun, F.; You, C.; Zhang, Y.-H.P. A high-energy-density sugar biobattery based on a synthetic enzymatic pathway. *Nat. Commun.* **2014**, *5*, 3026. [CrossRef]

50. Dudley, Q.M.; Nash, C.J.; Jewett, M.C. Cell-free biosynthesis of limonene using enzyme-enriched *Escherichia coli* lysates. *Synth. Biol.* **2019**, *4*, ysz003. [CrossRef]

51. Karim, A.S.; Jewett, M.C. Cell-free synthetic biology for pathway prototyping. *Methods Enzymol.* **2018**, *608*, 31–57.

52. Casini, A.; Chang, F.Y.; Eluere, R.; King, A.M.; Young, E.M.; Dudley, Q.M.; Karim, A.; Pratt, K.; Bristol, C.; Foget, A.; et al. A pressure test to make 10 molecules in 90 days: External evaluation of methods to engineer biology. *J. Am. Chem. Soc.* **2018**, *140*, 4302–4316. [CrossRef]

53. Karim, A.S.; Heggestad, J.T.; Crowe, S.A.; Jewett, M.C. Controlling cell-free metabolism through physiochemical perturbations. *Metab. Eng.* **2018**, *45*, 86–94. [CrossRef]

54. Bujara, M.; Schümperli, M.; Billerbeck, S.; Heinemann, M.; Panke, S. Exploiting cell-free systems: Implementation and debugging of a system of biotransformations. *Biotechnol. Bioeng.* **2010**, *106*, 376–389. [CrossRef]

55. Yi, T.; Lim, H.J.; Lee, S.J.; Lee, K.H.; Kim, D.M. Synthesis of (R,R)-2,3-butanediol from starch in a hybrid cell-free reaction system. *J. Ind. Eng. Chem.* **2018**, *67*, 231–235. [CrossRef]

56. Karim, A.S.; Jewett, M.C. A cell-free framework for rapid biosynthetic pathway prototyping and enzyme discovery. *Metab. Eng.* **2016**, *36*, 116–126. [CrossRef]

57. Lim, H.J.; Lee, K.H.; Kim, D.M. Rapid determination of effective folding agents by sequential cell-free protein synthesis. *Biochem. Eng. J.* **2018**, *138*, 106–110. [CrossRef]

58. Ahn, J.H.; Keum, J.W.; Kim, D.M. High-throughput, combinatorial engineering of initial codons for tunable expression of recombinant proteins. *J. Proteome Res.* **2008**, *7*, 2107–2113. [CrossRef]

59. Park, Y.J.; Lee, K.H.; Baek, M.S.; Kim, D.M. High-throughput engineering of initial coding regions for maximized production of recombinant proteins. *Biotechnol. Bioprocess Eng.* **2017**, *22*, 497–503. [CrossRef]

MDPI

St. Alban-Anlage 66

4052 Basel

Switzerland

Tel. +41 61 683 77 34

Fax +41 61 302 89 18

www.mdpi.com

Methods and Protocols Editorial Office

E-mail: mps@mdpi.com

www.mdpi.com/journal/mps